AF361889

Infrared Nanophotonics

Infrared Nanophotonics: Materials, Devices, and Applications

Editor

Tadaaki Nagao

MDPI • Basel • Beijing • Wuhan • Barcelona • Belgrade • Manchester • Tokyo • Cluj • Tianjin

Editor
Tadaaki Nagao
Photonics Nano-Engineering Group,
National Institute for Materials Science (NIMS)
Japan
Hokkaido University
Japan

Editorial Office
MDPI
St. Alban-Anlage 66
4052 Basel, Switzerland

This is a reprint of articles from the Special Issue published online in the open access journal *Micromachines* (ISSN 2072-666X) (available at: https://www.mdpi.com/journal/micromachines/special_issues/Infrared_Nanophotonics_Materials_Devices_Applications).

For citation purposes, cite each article independently as indicated on the article page online and as indicated below:

LastName, A.A.; LastName, B.B.; LastName, C.C. Article Title. *Journal Name* **Year**, *Volume Number*, Page Range.

ISBN 978-3-0365-0174-1 (Hbk)
ISBN 978-3-0365-0175-8 (PDF)

Contents

About the Editor

Tadaaki Nagao received his Dr. Sc. (physics) from Waseda University in 1995. He served as a Research Associate at Waseda University (1994), and at the University of Tokyo (1994–2001), and then an Associate Professor at Institute for Materials Research at Tohoku University (2001–2004). He joined National Institute for Materials Science in 2004 and currently serves as a Group Leader and MANA Principal Investigator. He also serves as a Professor in Condensed Matter Physics at Hokkaido University. His research interest lies in elucidating new electrical and optical functionalities at the surfaces of nanomaterials and applying them for radiative energy transfer/conversion.

Editorial

Editorial for the Special Issue "Infrared Nanophotonics: Materials, Devices and Applications"

Tadaaki Nagao [1,2]

[1]   International Center for Materials Nanoarchitectonics (MANA), National Institute for Materials Science, 1-1 Namiki, Tsukuba, Ibaraki 305-0044, Japan; NAGAO.Tadaaki@nims.go.jp

[2]   Department of Condensed Matter Physics, Graduate School of Science, Hokkaido University, Kita 8, Nishi 5, Kita-ku, Sapporo 060-0810, Japan

Received: 11 August 2020; Accepted: 19 August 2020; Published: 26 August 2020

Infrared light radiates from almost all the matter on earth and its strategic use will be an important issue for the enhancement of human life and the sustainable development of modern industry. Since it has frequency in the same region as phonons or molecular vibrations of materials, measuring its emission or absorption spectra helps us in characterizing and identifying the materials in a non-destructive manner. Meanwhile, if we can spectroscopically design the infrared emission by tuning the chemical composition or artificially controlling the nano- to mesoscale structures, it will have a great impact on industrial applications, such as in thermophotovoltaics, energy-saving drying furnaces, spectroscopic infrared light sources, and various types of infrared sensors.

In this Special Issue, recent studies by researchers who are working on MEMS-based infrared detectors, nanomaterial-based infrared detectors, thermal emitters, or fiber optics, have been contributed. Important topics of growing interests are the wavelength-selective infrared emitters and detectors where we can see rapid development in the field of nano-plasmonics and metamaterials, and we also collected contributions from narrow-band gap semiconductors.

This Special Issue collected 13 research papers, including one featured article. These are categorized as follows.

(1)   Infrared nano/micro devices based on lithographic techniques and MEMS structures.

Dao et al. have demonstrated a compact design for membrane-supported, wavelength-selective infrared (IR) bolometers [1]. The fabricated devices exhibit a wide resonance tunability in the mid-wavelength IR atmospheric window by changing the size of the resonator of the devices, evidencing that the concept of the proposed wavelength-selective IR bolometers is realizable. Dao et al. also experimentally studied the dark-field scattering spectral mapping of plasmonic resonance from the free-standing Al bowtie antenna arrays and correlated their strong nearfield enhancement with the sensing capability by means of surface-enhanced Raman spectroscopy [2]. Doan et al. reported a quad-wavelength hybrid plasmonic–pyroelectric detector that exhibited spectrally selective infrared detection at four wavelengths—3.3, 3.7, 4.1, and 4.5 μm [3]. The narrowband detection was achieved by coupling the incident infrared light to the resonant modes of the four different plasmonic perfect absorbers based on an Al-disk-array placed on an Al_2O_3–Al bilayer, exhibiting great possibilities for miniature multi-wavelength spectroscopic devices. Yoshino et al. developed, a simple process to mechanically fabricate ordered Au nanodot arrays that respond to nearinfrared light, and also reported the feasibility of its application to plasmonic sensors [4]. The developed nanoprocess utilizes direct mechanical cutting of Au film by single-crystal diamond blades and further thermal processing to tune the Au nanodot shape and their plasmon resonance. Sakurai et al. studied a tungsten-SiO_2-based metal insulator metal-structured metasurface for the thermal emitter of the thermophotovoltaic system [5]. The proposed emitter was fabricated by applying the photolithography method. The fabricated emitter has high emissivity in the visible to near-infrared region and shows excellent wavelength selectivity.

(2) Materials for infrared thermal emitters/absorbers and detectors based on compound semiconductors and their variants.

Ngo et al. reported the synthesis and demonstration of niobium-doped titanium dioxide for the application in plasmonic antenna and surface-enhanced infrared absorption [6]. The nanopatterns prepared by electron beam lithography, plasma etching/ashing processes showed well-defined antenna resonance as well as clear polarization/size dependence, which confirms that these materials are suitable for infrared plasmonic applications. Li et al. numerically studied the optical properties of hexagonal ITO nanodisk and nanohole arrays in the mid-infrared [7]. Field enhancement up to 10 times was observed in the simulated ITO nanostructures, and furthermore, they demonstrated the sensing of the surface phonon polariton from a 2-nm-thick SiO_2 layer under the ITO disk arrays. Chiu et al. examined the optical properties of alloys with noble metals (Au and Pt). The six different metals (Ir, Mo, Ni, Pb, Ta, and W) which possess good properties for heat resistance, stability, and magnetism were mixed with noble metals to improve the properties [8]. The optical properties were calculated by density functional theory and they were used for further investigations of the optical responses of alloy nanorods. The results show that the studied alloy nanorods have wavelength-selective properties and can be useful for infrared devices and systems. Zhai et al. reported a mid-wave infrared (MWIR) and long-wave infrared (LWIR) dual-band photodetector capable of voltage-controllable detection band selection [9]. The voltage-tunable dual-band photodetector is based on multiple stacks of sub-monolayer quantum dots (QDs) and self-assembled QDs. By changing the photodetector bias voltages, one can set the detection band to be MWIR, or LWIR, or both, with high photodetectivity and low crosstalk between the bands.

(3) Infrared-sensing applications using fiber and laser technology, and hyperspectral camera.

Inada et al. evaluated the performance of a fluorescent detection system in an extirpated pig stomach and a freshly resected human stomach and were able to successfully detect NIR fluorescence emitted from the clip in the stomach through the stomach wall by the irradiation of excitation light (λ: 808 nm) [10]. The proposed combined NIR light-emitting clip and laparoscopic fluorescent detection system could be useful in clinical practice for accurately identifying the location of a primary gastric tumor during laparoscopic surgery. Chen et al. reported a sensor system composed of a quantum cascade laser (4.65 μm excitation wavelength), and a compact multiple reflection cell with a light path length of 12 m for sensitively detecting trace CO gas [11]. The sensor adopted the long optical path differential absorption spectroscopy technique (LOP-DAST) and obtained the minimum detection limit (MDL) of 108 ppbv by comparing the residual difference between the measured spectrum and the Voigt theoretical spectrum. Mu et al. studied tunable diode laser absorption spectroscopy (TDLAS) combined with wavelength modulation spectroscopy (WMS) using an interband cascade laser for detecting a trace amount of C_2H_2 [12]. The data show that the minimum detection limit is as low as 1 ppbv at an integration time of 63 s, and capable of detecting a variety of gases by changing the wavelength of the laser. Kim et al. reported a novel real-time remote temperature estimation method by applying a deep-learning-based regression method to midwave infrared hyperspectral images [13]. They proposed a method for real-time remote temperature measurement with high accuracy with the proposed surface-temperature, deep convolutional neural network and a hyperspectral thermal camera.

We would like to thank all authors for submitting their papers; most of them kindly contributed to this Special Issue in response to our invitation. We would also like to acknowledge all the reviewers for dedicating their time and timely reviews to improve the quality of this Special Issue.

Conflicts of Interest: The author declares no conflict of interest.

References

1. Dao, T.D.; Doan, A.T.; Ishii, S.; Yokoyama, T.; Ørjan, H.S.; Ngo, D.H.; Ohki, T.; Ohi, A.; Wada, Y.; Niikura, C.; et al. MEMS-Based Wavelength-Selective Bolometers. *Micromachines* **2019**, *10*, 416. [CrossRef]

2.	Dao, T.D.; Hoang, C.V.; Nishio, N.; Yamamoto, N.; Ohi, A.; Nabatame, T.; Aono, M.; Nagao, T. Dark-Field Scattering and Local SERS Mapping from Plasmonic Aluminum Bowtie Antenna Array. *Micromachines* **2019**, *10*, 468. [CrossRef] [PubMed]

3.	Doan, A.T.; Yokoyama, T.; Dao, T.D.; Ishii, S.; Ohi, A.; Nabatame, T.; Wada, Y.; Maruyama, S.; Nagao, T. A MEMS-Based Quad-Wavelength Hybrid Plasmonic–Pyroelectric Infrared Detector. *Micromachines* **2019**, *10*, 413. [CrossRef] [PubMed]

4.	Yoshino, M.; Kubota, Y.; Nakagawa, Y.; Terano, M. Efficient Fabrication Process of Ordered Metal Nanodot Arrays for Infrared Plasmonic Sensor. *Micromachines* **2019**, *10*, 385. [CrossRef] [PubMed]

5.	Sakurai, A.; Matsuno, Y. Design and Fabrication of a Wavelength-Selective Near-Infrared Metasurface Emitter for a Thermophotovoltaic System. *Micromachines* **2019**, *10*, 157. [CrossRef] [PubMed]

6.	Ngo, H.D.; Chen, K.; Handegård, Ø.S.; Doan, A.T.; Ngo, T.D.; Dao, T.D.; Ikeda, N.; Ohi, A.; Nabatame, T.; Nagao, T. Nanoantenna Structure with Mid-Infrared Plasmonic Niobium-Doped Titanium Oxide. *Micromachines* **2020**, *11*, 23. [CrossRef] [PubMed]

7.	Li, Z.; Zhang, Z.; Chen, K. Indium–Tin–Oxide Nanostructures for Plasmon-Enhanced Infrared Spectroscopy: A Numerical Study. *Micromachines* **2019**, *10*, 241. [CrossRef] [PubMed]

8.	Chiu, M.-H.; Li, J.-H.; Nagao, T. Optical Properties of Au-Based and Pt-Based Alloys for Infrared Device Applications: A Combined First Principle and Electromagnetic Simulation Study. *Micromachines* **2019**, *10*, 73. [CrossRef] [PubMed]

9.	Zhai, Y.; Gu, G.; Lu, X. Voltage-Tunable Mid- and Long-Wavelength Dual-Band Infrared Photodetector Based on Hybrid Self-Assembled and Sub-Monolayer Quantum Dots. *Micromachines* **2019**, *10*, 4. [CrossRef] [PubMed]

10.	Inada, S.A.; Nakanishi, H.; Oda, M.; Mori, K.; Ito, A.; Hasegawa, J.; Misawa, K.; Fuchi, S. Development of a New Laparoscopic Detection System for Gastric Cancer Using Near-Infrared Light-Emitting Clips with Glass Phosphor. *Micromachines* **2019**, *10*, 81. [CrossRef] [PubMed]

11.	Chen, C.; Ren, Q.; Piao, H.; Wang, P.; Wang, Y. A Trace Carbon Monoxide Sensor Based on Differential Absorption Spectroscopy Using Mid-Infrared Quantum Cascade Laser. *Micromachines* **2018**, *9*, 670. [CrossRef] [PubMed]

12.	Mu, Y.; Hu, T.; Gong, H.; Ni, R.; Li, S. A Trace C_2H_2 Sensor Based on an Absorption Spectrum Technique Using a Mid-Infrared Interband Cascade Laser. *Micromachines* **2018**, *9*, 530. [CrossRef] [PubMed]

13.	Kim, S.; Kim, J.; Lee, J.; Ahn, J. Midwave FTIR-Based Remote Surface Temperature Estimation Using a Deep Convolutional Neural Network in a Dynamic Weather Environment. *Micromachines* **2018**, *9*, 495. [CrossRef] [PubMed]

Article

Nanoantenna Structure with Mid-Infrared Plasmonic Niobium-Doped Titanium Oxide

Hai Dang Ngo [1,2], Kai Chen [1,3], Ørjan S. Handegård [1,2], Anh Tung Doan [1,2], Thien Duc Ngo [1,2], Thang Duy Dao [1], Naoki Ikeda [1,4], Akihiko Ohi [1,4], Toshihide Nabatame [1,4] and Tadaaki Nagao [1,2,*]

[1]   International Center for Materials Nanoarchitectonics, National Institute for Materials Science, Tsukuba 305-0044, Japan; NGO.HaiDang@nims.go.jp (H.D.N.); kaichen@jnu.edu.cn (K.C.); HANDEGARD.Orjansele@nims.go.jp (Ø.S.H.); DOAN.TungAnh@nims.go.jp (A.T.D.); DUCTHIEN.Ngo@nims.go.jp (T.D.N.); katsiusa@gmail.com (T.D.D.); IKEDA.Naoki@nims.go.jp (N.I.); OHI.Akihiko@nims.go.jp (A.O.); NABATAME.Toshihide@nims.go.jp (T.N.)
[2]   Department of Condensed Matter Physics, Graduate school of Science, Hokkaido University, Kita-10 Nishi-8 Kita-ku, Sapporo 060-0810, Japan
[3]   Institute of Photonics Technology, Jinan University, Guangzhou 510632, China
[4]   Nanotechnology Innovation Station, National Institute for Materials Science, Tsukuba 305-0044, Japan
*   Correspondence: NAGAO.Tadaaki@nims.go.jp

Received: 17 November 2019; Accepted: 23 December 2019; Published: 24 December 2019

Abstract: Among conductive oxide materials, niobium doped titanium dioxide has recently emerged as a stimulating and promising contestant for numerous applications. With carrier concentration tunability, high thermal stability, mechanical and environmental robustness, this is a material-of-choice for infrared plasmonics, which can substitute indium tin oxide (ITO). In this report, to illustrate great advantages of this material, we describe successful fabrication and characterization of niobium doped titanium oxide nanoantenna arrays aiming at surface-enhanced infrared absorption spectroscopy. The niobium doped titanium oxide film was deposited with co-sputtering method. Then the nanopatterned arrays were prepared by electron beam lithography combined with plasma etching and oxygen plasma ashing processes. The relative transmittance of the nanostrip and nanodisk antenna arrays was evaluated with Fourier transform infrared spectroscopy. Polarization dependence of surface plasmon resonances on incident light was examined confirming good agreements with calculations. Simulated spectra also present red-shift as length, width or diameter of the nanostructures increase, as predicted by classical antenna theory.

Keywords: nanoantenna; niobium-doped titanium oxide; mid-infrared plasmonics

---

1. Introduction

In recent times, infrared nano-plasmonics shifts its interest from common noble metals or compounds like Au, Pt, Ag or TiN [1] to transparent conductive oxides as materials of choice [2]—namely, indium tin oxide [3–9], fluorine doped tin oxide [10,11], aluminum (or gallium) doped zinc oxide [12–17]. Extensive research on those materials aims at both basic properties in infrared region as well as different applications such as wavelength-selective perfect absorbers and emitters [6,18,19] interacting plasmonic nano-particles [20–23], optical meta-surfaces [19], and active tunable plasmonic devices [7,19,24]. Compared to those oxide materials, niobium-doped titanium oxide (TiO_2:Nb) exhibits notable electrical and optical properties. The so-called name "transparent metal" comes from high transparency in visible light and noticeably low electrical resistivity. It also has great thermal durability, high surface smoothness, mechanical robustness, humid environment endurance and other advantages, just to name a few [25–32].

In this report, we demonstrate the use of TiO_2:Nb in nanostrip and nanodisk antenna arrays for infrared plasmonic devices such as surface-enhanced infrared absorption spectroscopy (SEIRA). The nanopatterned arrays were fabricated using electron beam lithography combined with plasma etching and oxygen plasma ashing processes. The fabricated nanostrip and nanodisk antenna structures of TiO_2:Nb were sequentially subjected to linearly polarized infrared light in relative transmittance measurement with Fourier transform infrared spectroscopy. Resulted plasmon resonances exposed polarization dependence and shifted to longer wavelengths as the length, width of nanostrips, or diameter of nanodisks expanded. Experiment and simulation results are in good concordance.

2. Materials and Methods

Fabrication process of niobium-doped titanium dioxide nanoantenna is demonstrated generally in Figure 1.

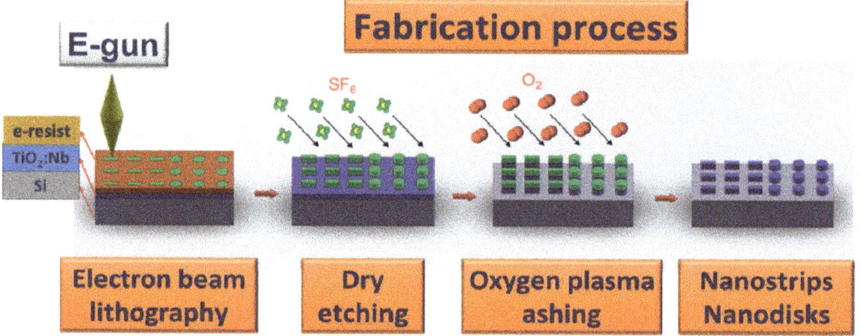

Figure 1. Fabrication process of Nb-doped TiO_2 nanoantenna (from left to right) consists of electron beam lithography, development, dry etching with SF_6 gas, and oxygen plasma ashing. The electron resist layer (**brown**) was coated on TiO_2:Nb thin film (**blue**), which was deposited on silicon substrate (**grey**). The remained electron resist layer (**green**) after lithography process helped to define the nanostrip and nanodisk of antenna structures.

At first, the Nb-doped TiO_2 film was co-sputtered at room temperature with TiO_2 ceramic and Nb metallic targets. Both silicon (001) and borosilicate glass were used simultaneously as sputtering substrates in every deposition. Radio frequency (RF) and direct current (DC) sputtering methods were used for TiO_2 and Nb targets, respectively. The sputtering chamber (!-Miller, Shibaura Mechatronics Corporation, Yokohama, Japan) was initially evacuated down to 4.0×10^{-5} Pa as base pressure, then 19 sccm of argon and 1 sccm of oxygen gas flow was introduced into the chamber to create working pressure of about 0.3 Pa. Sputtering power was set at 200 W (RF) and 20 W (DC) for TiO_2 and Nb targets, respectively. After one hour of deposition, 80-nm-thick film was obtained. Then, the as-deposited films went through vacuum thermal annealing at 600 °C for 1 h. Optical properties of TiO_2:Nb thin films were characterized with Ellipsometry (SENTECH, SE 850 DUV and SENDIRA, SENTECH Instruments GmbH, Berlin, Germany) from deep ultraviolet to far infrared region.

Consequently, the samples were spin coated with negative resist NBE-22A for electron beam lithography (at 2000 rounds per minute in 60 s) and baked with hot plate (110 °C in 5 min). Electron beam lithography (ELIONIX, ELS-7500EX, ELIONIX INC, Tokyo, Japan) was used to write desired pattern onto the resist layer (Figure 1). After pattern writing process, samples were subjected to post-baking (110 °C in 5 min), developing with NMD-3 in 60 s and rinsing with iso-propyl alcohol (IPA), and finally blow-drying with nitrogen gas gun.

Reactive ion etching (ULVAC, CE-300I, SF_6 gas, 0.5 Pa, 100 W power) and oxygen plasma ashing (Mory PB-600, 300 W power in 15 min) steps were carried out to remove unnecessary surrounding TiO_2:Nb nanostrips or nanodisks and wash away the remaining electron resist.

Shape and morphology of strip and disk nanostructures were investigated with scanning electron microscopy (SEM, Hitachi, SU-8400, Hitachi High-Technologies Corporation, Tokyo, Japan) and Atomic force microscope (AFM, Nanoscope 5, Bruker Corporation, Billerica, MA, USA). The silicon tips SI-DF20 (Hitachi High Technologies, Tokyo, Japan) were used at tapping mode.

Fourier transform infrared spectroscopy (FTIR, Thermo Scientific Nicolet iS50, Waltham, MA, USA) was used to assess the resonance features of nanostructures with different linear polarization of incident light, i.e., electric field vector was either parallel to length ($\vec{E}_{\parallel}$) or width ($\vec{E}_{\perp}$)of nanostrips, as illustrated in Figure 2. Unpolarized incident light was used for nanodisks.

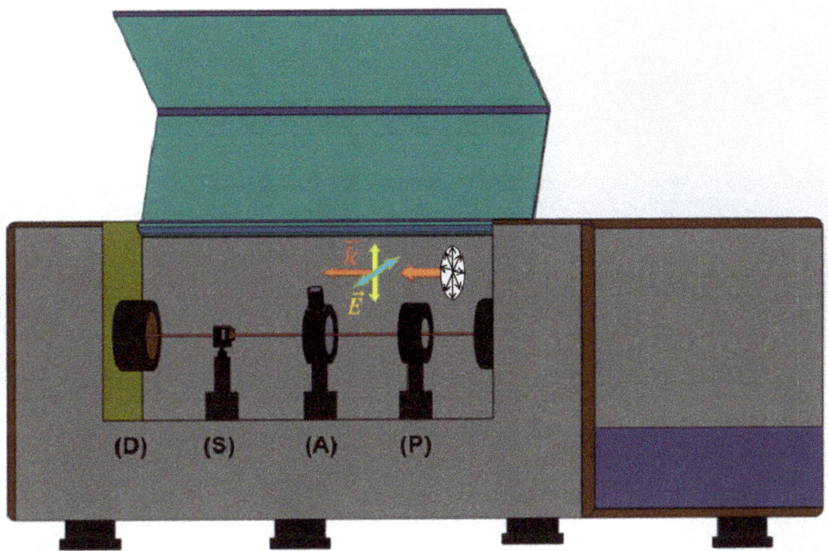

Figure 2. Transmission measurement setup with Fourier transform infrared spectrometer. Unpolarized incident infrared light (right side) passed through polarizer (**P**), aperture (**A**) and then exposed to nanostrip sample (**S**) before entering the detector (**D**). In case of nanodisk sample, polarizer (**P**) was removed and unpolarized infrared light was used.

3. Results and Discussion

Dielectric functions (ε_1, ε_2) of co-sputtered Nb-doped TiO_2 films on silicon substrate (001) were characterized with spectroscopic ellipsometry at three different reflected angles of 60°, 65°, and 70° from ultraviolet (0.3 µm) to far infrared (25 µm) region (Figure 3).

Real part of the dielectric function (ε_1) exhibited a cross-over point in the mid-infrared region at approximately 8.5 µm. By increasing the Nb sputtering power, carrier concentration can be tuned, and Fermi level can easily be shifted toward conduction band. This in turn helps to adjust the cross-over point back and forth in the infrared region, whereas for conventional plasmonic metals such continuous tunability is not possible. Furthermore, this material also has great thermal durability, high surface smoothness, mechanical robustness, as well as photocatalytic activity and humid environment endurance. Combining altogether, TiO_2:Nb film proves itself an excellent candidate for diverse infrared plasmonic applications.

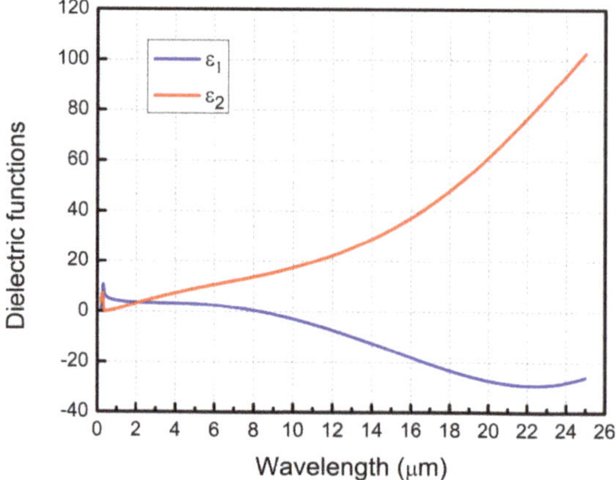

Figure 3. Experimental dielectric functions of TiO_2:Nb films on silicon substrate. Real (ε_1) and imaginary (ε_2) parts in blue and red curves, respectively.

To realize the plasmonic property of co-sputtered TiO_2:Nb films in infrared region, nanostrips of different sizes were simulated with electromagnetic solver to check for the antenna resonance. The simulated parameters were transferred into lithographic patterning and fabrication, and then characterized by the IR spectroscopy. Figure 4a showed general designed nanostrip pattern used for Finite-Difference Time-Domain method (FDTD, Rsoft, Synopsis) simulation as well as electron beam lithography. In simulation, the lengths (L) of the strips were set from 650 to 750 nm while the width values (W) varied from 400 to 600 nm.

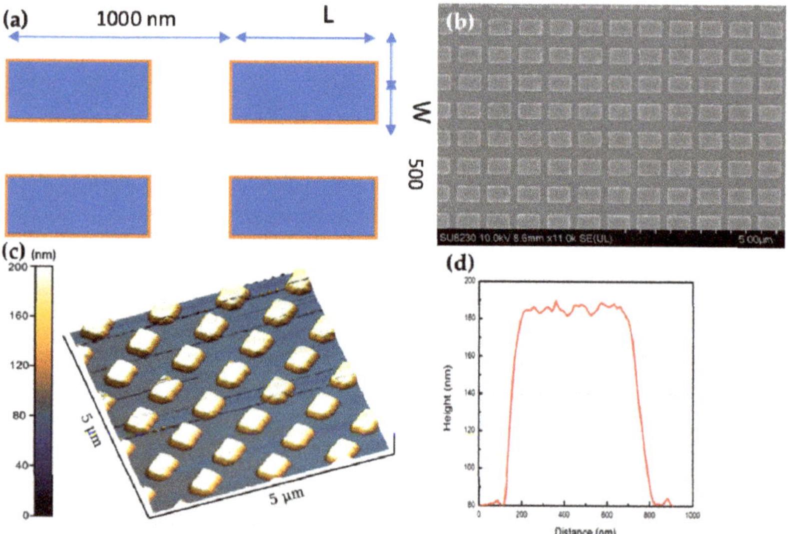

Figure 4. Design patterns of TiO_2:Nb antenna for simulation and electron beam lithography (**a**). Horizontal periodicity is 1000 nm. Strip width (W) is about 500 nm. SEM image of TiO_2:Nb nanostrips (**b**). Three-dimensional AFM image (5 μm × 5 μm) of TiO_2:Nb nanostrips (**c**). Height profile of a single nanostrip (**d**).

SEM and AFM images (Figure 4b,c) show periodic rectangular nanostrips, which are similar to intended design pattern (Figure 4a). The average strip surface roughness of 2.3 nm confirms that the etching and ashing processes did not cause considerable damage.

The smallest interval between two consecutive fabricated strips is approximately 300 nm. As illustrated in Figures 5d and 6d, relative transmittance exhibits strong fundamental antennalike resonances as polarized electric field vector is parallel to nanostrip length or width. Both measured and simulated spectra show good agreement. Calculated spectra demonstrate orderly shift of resonances toward longer wavelengths (black dash arrow) and reveal the improvement in quality factor of nanoantenna (sharper and deeper resonances) as length or width of nanostrips increases, as in the case of metal nanowires [32]. FDTD simulation results along x–y and x–z planes (Figure 5d,e and Figure 6d,e) also show strong electric field confinement and enhancement at the edges of nanostrips. This proves the antennalike dipole resonance of the structures.

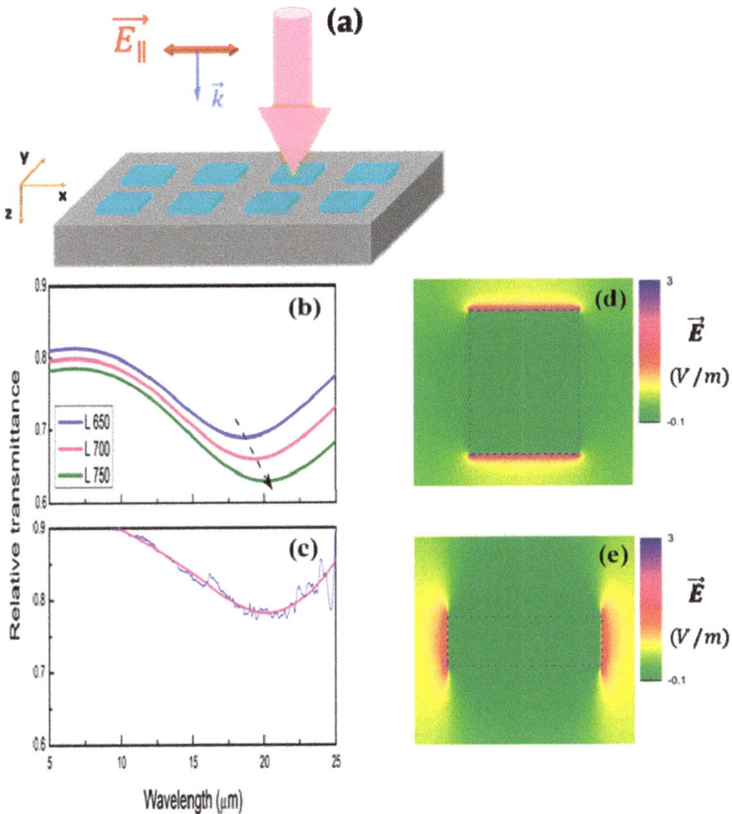

Figure 5. Schematic illustration of the exposure of polarized infrared light with electric field vector (k) parallel to length of nanostrips ($\vec{E}_{\parallel}$) (**a**). Simulation (**b**) and FTIR relative transmittance (**c**) of nanostrips with ($\vec{E}_{\parallel}$) polarized incident light. Lengths of nanostrips (L) in simulation were set at 650, 700, and 750 nm, corresponding to blue, pink, and green curves (**b**). Measured and fitted lines were drawn in blue and pink, respectively (L × W ~ 750 nm × 500 nm) (**c**). Electric field distribution along x–y and x–z planes of strips (**d**, **e**).

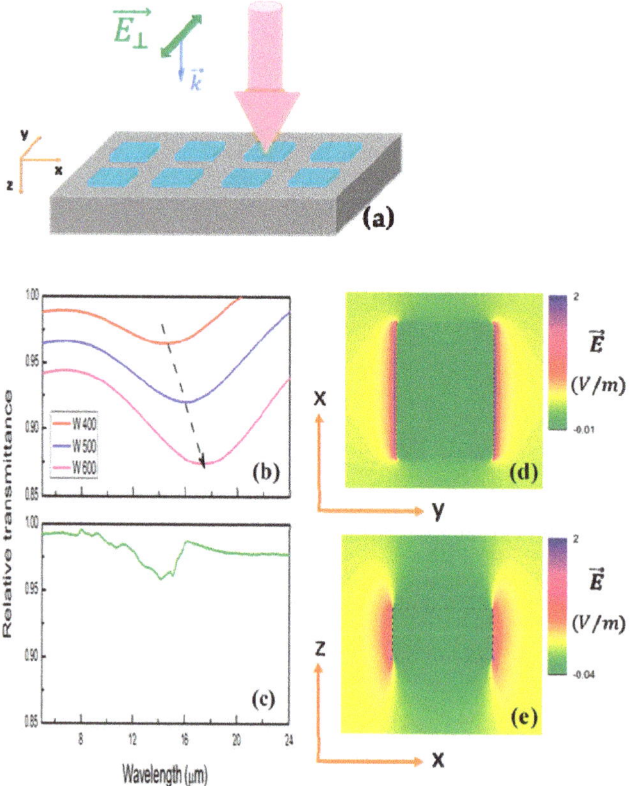

Figure 6. Schematic illustration of polarized infrared light with electric field vector parallel to the longer axis of nanostrips ($\vec{E}_\perp$) (**a**). Simulation (**b**) and FTIR relative transmittance (**c**) of nanostrips with ($\vec{E}_\perp$) polarized incident light. Widths of nanostrips (W) in simulation were set at 400, 500, and 600 nm, corresponding to red, blue, and pink curves (**b**). Measured spectrum was plotted in green line (L × W ~ 700 nm × 450 nm) (**c**). Electric field distribution along x-y and x-z planes of strips (**d**, **e**).

General designed nanodisk pattern for FDTD calculation and electron beam lithography was shown in Figure 7a. Diameters of nanodisks (D) were set from 600 to 800 nm.

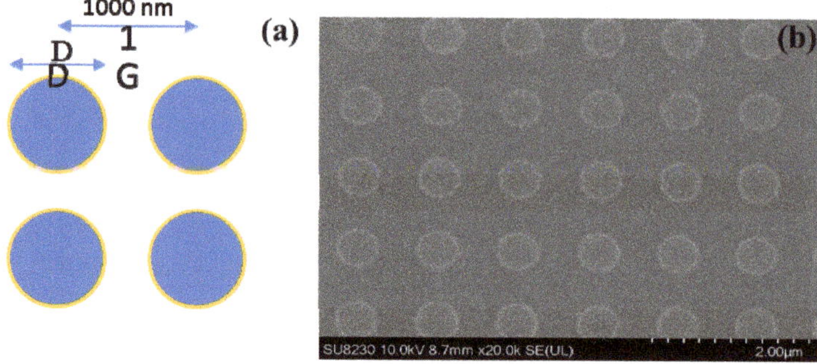

Figure 7. Design patterns of TiO_2:Nb nanodisk antennas for calculation and fabrication. Periodicity is 1000 nm (**a**). SEM image of TiO_2:Nb nanodisk structure (**b**).

In simulated results, systematic increase of resonance frequency with diameter (D) can be easily observed. Sharper resonances can be expected if disk diameter keeps expanding. Experimental relative transmittance spectrum is in consonance with calculated lines with strong fundamental antennalike resonances under any linear polarized infrared light parallel to the disk surface (Figure 8d). Quality factor of these nanodisk antennae can be further improved when the diameter D increases since the surface plasmon resonance assumes more photonic nature and the inherent loss of the material becomes less. While resonant frequencies show good matching between simulation and experiment curves in x axis; there is some difference (~10% to 15%) in relative transmittance (y-axis). This can be explained by two factors: fabrication tolerance and etchant effect. Perfect cuboids were used in simulation, but the obtained ones were not perfectly flat at edges, as shown in Figure 4c,d. Furthermore, electron carrier concentration was decreased from 1.12×10^{21} cm^{-3} for pristine film to 6.4×10^{20} cm^{-3} for remained one. SF$_6$ gas may introduce some electron-trapping fluorine ions on surface of nanostructure after etching.

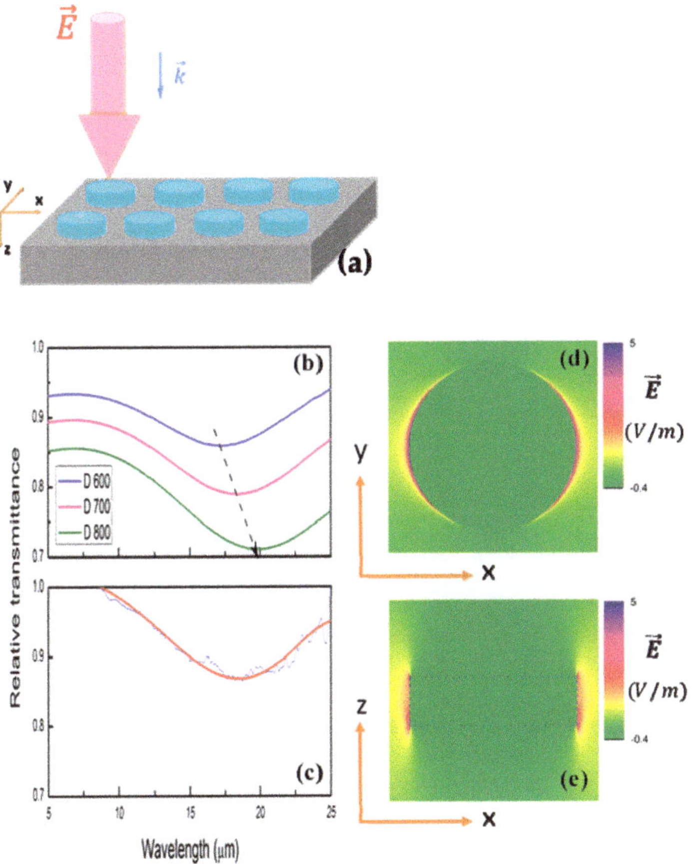

Figure 8. Schematic diagram of unpolarized infrared light on nanodisk antenna structures (**a**). Simulation (**b**) and FTIR relative transmittance (**c**) of nanodisk with unpolarized incident light. Diameters of nanodisks (D) in simulation were set at 600, 700, and 800 nm, corresponding to blue, pink, and green curves. Measured and fitted lines were drawn in blue and red (D ~750 nm) (**c**). Electric field distribution along x–y and x–z planes (**d**, **e**).

4. Conclusions

We succeeded in designing and fabricating antenna arrays of nanostrips and nanodisks based on Nb doped TiO_2. Relative infrared transmittance in calculated and experimental results are in good accordance. Spectral features exhibit red-shift in their peak position as the sizes of strips and disks increase. In the same tendency, quality factor of nanostrips and nanodisks can also be improved if structure dimension can be further extended.

Author Contributions: Conceptualization, T.N.; methodology, H.D.N., K.C., A.O. and N.I.; software, A.T.D., T.D.N.; validation, and T.D.D.; formal analysis, H.D.N.; investigation, H.D.N., K.C.; data curation, H.D.N.; writing—original draft preparation, H.D.N.; writing—review and editing, H.D.N. and Ø.S.H.; visualization, H.D.N., A.T.D.; supervision, T.N.; project administration, T.N.; funding acquisition, T.N. All authors have read and agreed to the published version of the manuscript.

Funding: T.N. acknowledges the support from CREST under the project of "Phase Interface Science for Highly Efficient Energy Utilization" (JPMJCR13C3, Japan Science and Technology Agency). T.N. also acknowledge the support from JSPS KAKENHI Grant Numbers 16H06364.

Acknowledgments: We thank Tomoko Ohki and Katsumi Ohno at Nanotechnology Innovation Station, National Institute for Materials Science, for their kind help.

Conflicts of Interest: There are no conflicts of interest to declare.

References

1. Sugavaneshwar, R.P.; Ishii, S.; Dao, T.D.; Ohi, A.; Nabatame, T.; Nagao, T. Fabrication of Highly Metallic TiN Films by Pulsed Laser Deposition Method for Plasmonic Applications. *ACS Photonics* **2018**, *5*, 814–819. [CrossRef]

2. Kim, J.; Naik, G.V.; Emani, N.K.; Guler, U.; Boltasseva, A. Plasmonic Resonances in Nanostructured Transparent Conducting Oxide Films. *IEEE J. Sel. Top. Quantum Electron.* **2013**, *19*, 4601907.

3. Wang, Y.; Overvig, A.C.; Shrestha, S.; Zhang, R.; Wang, R.; Yu, N.; Dal Negro, L. Tunability of indium tin oxide materials for mid-infrared plasmonics applications. *Opt. Mater. Express* **2017**, *7*, 2727–2739. [CrossRef]

4. Li, S.Q.; Guo, P.; Zhang, L.; Zhou, W.; Odom, T.W.; Seideman, T.; Ketterson, J.B.; Chang, R.P.H. Infrared Plasmonics with Indium–Tin-Oxide Nanorod Arrays. *ACS Nano* **2011**, *5*, 9161–9170. [CrossRef] [PubMed]

5. Rhodes, C.; Cerruti, M.; Efremenko, A.; Losego, M.; Aspnes, D.E.; Maria, J.-P.; Franzen, S. Dependence of plasmon polaritons on the thickness of indium tin oxide thin films. *J. Appl. Phys.* **2008**, *103*, 93108. [CrossRef]

6. Tamanai, A.; Dao, T.D.; Sendner, M.; Nagao, T.; Pucci, A. Mid-infrared optical and electrical properties of indium tin oxide films. *Phys. Status Solidi* **2017**, *214*, 1600467. [CrossRef]

7. Guo, P.; Schaller, R.D.; Ketterson, J.B.; Chang, R.P.H. Ultrafast switching of tunable infrared plasmons in indium tin oxide nanorod arrays with large absolute amplitude. *Nat. Photonics* **2016**, *10*, 267. [CrossRef]

8. Fang, X.; Mak, C.L.; Dai, J.; Li, K.; Ye, H.; Leung, C.W. ITO/Au/ITO Sandwich Structure for Near-Infrared Plasmonics. *ACS Appl. Mater. Interfaces* **2014**, *6*, 15743–15752. [CrossRef]

9. Krasilnikova Sytchkova, A.; Grilli, M.L.; Boycheva, S.; Piegari, A. Optical, electrical, structural and microstructural characteristics of rf sputtered ITO films developed for art protection coatings. *Appl. Phys. A* **2007**, *89*, 63–72. [CrossRef]

10. Khalilzadeh-Rezaie, F.; Oladeji, I.O.; Cleary, J.W.; Nader, N.; Nath, J.; Rezadad, I.; Peale, R.E. Fluorine-doped tin oxides for mid-infrared plasmonics. *Opt. Mater. Express* **2015**, *5*, 2184–2192. [CrossRef]

11. Dominici, L.; Michelotti, F.; Brown, T.M.; Reale, A.; Carlo, A. Di Plasmon polaritons in the near infrared on fluorine doped tin oxide films. *Opt. Express* **2009**, *17*, 10155–10167. [CrossRef] [PubMed]

12. Naik, G.V.; Liu, J.; Kildishev, A.V.; Shalaev, V.M.; Boltasseva, A. Demonstration of Al:ZnO as a plasmonic component for near-infrared metamaterials. *Proc. Natl. Acad. Sci. USA* **2012**, *109*, 8834–8838. [CrossRef] [PubMed]

13. Buonsanti, R.; Llordes, A.; Aloni, S.; Helms, B.A.; Milliron, D.J. Tunable Infrared Absorption and Visible Transparency of Colloidal Aluminum-Doped Zinc Oxide Nanocrystals. *Nano Lett.* **2011**, *11*, 4706–4710. [CrossRef] [PubMed]

14. Kim, J.; Naik, G.V.; Gavrilenko, A.V.; Dondapati, K.; Gavrilenko, V.I.; Prokes, S.M.; Glembocki, O.J.; Shalaev, V.M.; Boltasseva, A. Optical Properties of Gallium-Doped Zinc Oxide—A Low-Loss Plasmonic Material: First-Principles Theory and Experiment. *Phys. Rev. X* **2013**, *3*, 041037. [CrossRef]

15. Hendrickson, J.R.; Vangala, S.; Nader, N.; Leedy, K.; Guo, J.; Cleary, J.W. Plasmon resonance and perfect light absorption in subwavelength trench arrays etched in gallium-doped zinc oxide film. *Appl. Phys. Lett.* **2015**, *107*, 191906. [CrossRef]
16. Sachet, E.; Losego, M.D.; Guske, J.; Franzen, S.; Maria, J.-P. Mid-infrared surface plasmon resonance in zinc oxide semiconductor thin films. *Appl. Phys. Lett.* **2013**, *102*, 051111. [CrossRef]
17. Pradhan, A.K.; Mundle, R.M.; Santiago, K.; Skuza, J.R.; Xiao, B.; Song, K.D.; Bahoura, M.; Cheaito, R.; Hopkins, P.E. Extreme tunability in aluminum doped Zinc Oxide plasmonic materials for near-infrared applications. *Sci. Rep.* **2014**, *4*, 6415. [CrossRef]
18. Kesim, Y.E.; Battal, E.; Okyay, A.K. Plasmonic materials based on ZnO films and their potential for developing broadband middle-infrared absorbers. *AIP Adv.* **2014**, *4*, 077106. [CrossRef]
19. Dao, T.D.; Doan, A.T.; Ngo, D.H.; Chen, K.; Ishii, S.; Tamanai, A.; Nagao, T. Selective thermal emitters with infrared plasmonic indium tin oxide working in the atmosphere. *Opt. Mater. Express* **2019**, *9*, 2534–2544. [CrossRef]
20. Kanehara, M.; Koike, H.; Yoshinaga, T.; Teranishi, T. Indium Tin Oxide Nanoparticles with Compositionally Tunable Surface Plasmon Resonance Frequencies in the Near-IR Region. *J. Am. Chem. Soc.* **2009**, *131*, 17736–17737. [CrossRef]
21. Wang, T.; Radovanovic, P.V. Free Electron Concentration in Colloidal Indium Tin Oxide Nanocrystals Determined by Their Size and Structure. *J. Phys. Chem. C* **2011**, *115*, 406–413. [CrossRef]
22. Xi, M.; Reinhard, B.M. Localized Surface Plasmon Coupling between Mid-IR-Resonant ITO Nanocrystals. *J. Phys. Chem. C* **2018**, *122*, 5698–5704. [CrossRef] [PubMed]
23. Chen, K.; Guo, P.; Dao, T.D.; Li, S.-Q.; Ishii, S.; Nagao, T.; Chang, R.P.H. Protein-Functionalized Indium-Tin Oxide Nanoantenna Arrays for Selective Infrared Biosensing. *Adv. Opt. Mater.* **2017**, *5*, 1700091. [CrossRef]
24. Liu, X.; Kang, J.-H.; Yuan, H.; Park, J.; Kim, S.J.; Cui, Y.; Hwang, H.Y.; Brongersma, M.L. Electrical tuning of a quantum plasmonic resonance. *Nat. Nanotechnol.* **2017**, *12*, 866. [CrossRef]
25. Furubayashi, Y.; Hitosugi, T.; Yamamoto, Y.; Inaba, K.; Kinoda, G.; Hirose, Y.; Shimada, T.; Hasegawa, T. A transparent metal: Nb-doped anatase TiO_2. *Appl. Phys. Lett.* **2005**, *86*, 252101. [CrossRef]
26. Zhang, S.X.; Kundaliya, D.C.; Yu, W.; Dhar, S.; Young, S.Y.; Salamanca-Riba, L.G.; Ogale, S.B.; Vispute, R.D.; Venkatesan, T. Niobium doped TiO_2: Intrinsic transparent metallic anatase versus highly resistive rutile phase. *J. Appl. Phys.* **2007**, *102*, 013701. [CrossRef]
27. Baumard, J.F.; Tani, E. Electrical conductivity and charge compensation in Nb doped TiO_2 rutile. *J. Chem. Phys.* **1977**, *67*, 857–860. [CrossRef]
28. Lee, H.-Y.; Robertson, J. Doping and compensation in Nb-doped anatase and rutile TiO_2. *J. Appl. Phys.* **2013**, *113*, 213706. [CrossRef]
29. Arbiol, J.; Cerdà, J.; Dezanneau, G.; Cirera, A.; Peiró, F.; Cornet, A.; Morante, J.R. Effects of Nb doping on the TiO_2 anatase-to-rutile phase transition. *J. Appl. Phys.* **2002**, *92*, 853–861. [CrossRef]
30. Yamada, N.; Hitosugi, T.; Kasai, J.; Hoang, N.L.H.; Nakao, S.; Hirose, Y.; Shimada, T.; Hasegawa, T. Direct growth of transparent conducting Nb-doped anatase TiO_2 polycrystalline films on glass. *J. Appl. Phys.* **2009**, *105*, 123702. [CrossRef]
31. Jaćimović, J.; Gaál, R.; Magrez, A.; Piatek, J.; Forró, L.; Nakao, S.; Hirose, Y.; Hasegawa, T. Low temperature resistivity, thermoelectricity, and power factor of Nb doped anatase TiO_2. *Appl. Phys. Lett.* **2013**, *102*, 013901.
32. Neubrech, F.; Kolb, T.; Lovrincic, R.; Fahsold, G.; Pucci, A.; Aizpurua, J.; Cornelius, T.W.; Toimil-Molares, M.E.; Neumann, R.; Karim, S. Resonances of individual metal nanowires in the infrared. *Appl. Phys. Lett.* **2006**, *89*, 253104. [CrossRef]

Article

Dark-Field Scattering and Local SERS Mapping from Plasmonic Aluminum Bowtie Antenna Array

Thang Duy Dao [1,2], Chung Vu Hoang [3,4,*], Natsuki Nishio [5], Naoki Yamamoto [5], Akihiko Ohi [1], Toshihide Nabatame [1], Masakazu Aono [1] and Tadaaki Nagao [1,6,*]

[1] International Center for Materials Nanoarchitectonics (MANA), National Institute for Materials Science (NIMS), 1-1 Namiki, Tsukuba 305-0044, Japan
[2] Graduate School of Materials Science, Nara Institute of Science and Technology, 8916-5 Takayama, Ikoma, Nara 630-0192, Japan
[3] Institute of Materials Science (IMS), Vietnam Academy of Science and Technology (VAST), 18 Hoang Quoc Viet street, Hanoi 100000, Vietnam
[4] Institute of Theoretical and Applied Research (ITAR), Duy Tan University, 1 Phung Chi Kien Street, Hanoi 100000, Vietnam
[5] Physics Department, Tokyo Institute of Technology, Meguro-ku, Tokyo 152-8551, Japan
[6] Department of Condensed Matter Physics, Graduate School of Science, Hokkaido University, Kita-10 Nishi-8 Kita-ku, Sapporo 060-0810, Japan
* Correspondence: chunghv@ims.vast.ac.vn (C.V.H.); nagao.tadaaki@nims.go.jp (T.N.); Tel.: +84-24-375-64129 (C.V.H.); +81-29-860-4709 (T.N.)

Received: 22 April 2019; Accepted: 11 July 2019; Published: 13 July 2019

Abstract: On the search for the practical plasmonic materials beyond noble metals, aluminum has been emerging as a favorable candidate as it is abundant and offers the possibility of tailoring the plasmonic resonance spanning from ultra-violet to the infrared range. In this letter, in combination with the numerical electromagnetic simulations, we experimentally study the dark-field scattering spectral mapping of plasmonic resonance from the free-standing Al bowtie antenna arrays and correlate their strong nearfield enhancement with the sensing capability by means of surface-enhanced Raman spectroscopy. The spatial matching of plasmonic and Raman mapping puts another step to realize a very promising application of free-standing Al bowtie antennas for plasmonic sensing.

Keywords: aluminum; plasmonics; dark-field scattering; SERS; bow-tie antenna

---

1. Introduction

Alternative plasmonic materials beyond gold (Au) and silver (Ag), including transition metal nitrides and metal carbides [1–4], heavily-doped semiconductors [5–9], as well as graphene [10–15], have been widely investigated recently owing to their controllability of optical properties via chemical or physical doping processes. Being the third most abundant element on earth, after silicon and oxygen, aluminum (Al) has also been used widely in industry. In the past few years, aluminum have attracted tremendous interest in the field of plasmonics as it is a promising alternative replacement for noble plasmonic metals [16–20]. Having an ultrathin native oxide Al_2O_3 layer (2 nm–4 nm), Al emerges as a favorable material in the field of plasmonic sensing with the detection mechanism mainly based on the electromagnetic field enhancement rather than the charge transfer effect [16,17,21,22]. Compared to Au and Ag whose interband transitions are at ~2.3 eV and ~3.9 eV [23], respectively; Al has a lower interband transition energy of about ~1.5 eV. Aluminum displays a plasma energy of 15.3 eV thatis higher than that of Au and Ag, and, its energy loss function $Im\left(\frac{-1}{\varepsilon}\right)$, wherein ε is the dielectric function of the material, offers no maxima of dielectric loss in the range from UV to VIS, thus making Al an excellent Drude metal in this range [24–27]. In terms of the optical absorption, the absorption efficiency of Al nanoparticles is greater than that of Au and Ag, and is slightly smaller than that of alkali metals [28]. Therefore, plasmonic resonance of Al nanoantennas can be flexibly tuned over the UV-VIS to the IR region, just by adopting suitable nanostructured architectures [29–34].

In the past decades, plasmon-enhanced vibrational spectroscopic sensing including surface-enhanced Raman scattering (SERS) [35,36] and surface-enhanced infrared absorption spectroscopy (SEIRA) [37–39] have showed great advantage over the conventional spectroscopic methods in the trace molecular detection, especially for monolayer and single molecule via specific surface chemical functionalization [40–45]. In the field of SERS, many plasmonic antennas have been proposed for SERS substrates including plasmonic nanospheres [40,46], nanorods [47,48], nanotriangles [17,49], particularly the structures having narrow gaps such as aggregate nanoparticles [50], nanoparticle dimers [46,51], nanoclusters [51,52], and nano-bowties [22,53,54]. Among them, plasmonic nano-bowties have attracted much attention because of its largest nearfield enhancement that could be achieved at the nanogaps between the two triangles [22,53]. On the other hand, the Raman cross section is increased in the short wavelength(λ) region (UV-NIR) as a function of λ^{-4}, however, engineering the resonance of bowtie antennas in the UV-NIR region using Au and Ag is impossible due to their small plasma frequencies and the fabrication limit. Therefore, utilizing plasmonic Al for nanobowtie antennas can realize short-wavelength resonant antennas (500 nm–700 nm) for SERS applications.

In this letter, we numerically and experimentally demonstrated free-standing Al bowtie antennas for SERS study. As the group velocity of the "slow light" formed by the coupling of plasmon and light is screened by the dielectric property of the surrounding media, thus, having an equality of the group velocity by a homogeneous surrounded dielectric medium between the upper and lower metal/dielectric interfaces would give a better homogeneity of the distribution of the electromagnetic field [54]. Compared to the previous Al bowtie antennas for SERS that the Al bowties was placed directly on a fused silica substrate [22], in this work, by introducing a SiO_2 post under each Al triangle, the nanogap area becomes free from the dielectric screening and thus the induced field enhancement at the nanogaps located between two plasmonic Al triangles could be increased and occupy a larger available volume for accommodating analyte molecules. We investigated the dark-field scattering mapping spectra of the plasmonic free-standing Al bowtie antennas, for correlating the obtained resonance with their sensing capability by the surface-enhanced Raman spectroscopy (SERS) of organic molecules. Supported by the numerical simulations, we showed that the free-standing Al bowtie arrays could be a powerful antenna platform for SERS. Our result also paves a way for the possible applications of the oxide-coated Al bowtie array serving as strong near-field optical antennas for SERS as well as other plasmon-enhanced spectroscopic devices.

2. Materials and Methods

The free-standing Al bowtie nanoantennas were fabricated by a standard procedure using electron beam (EB) lithography, lift-off, and reactive-ion etching (RIE) processes (Figure 1a). Prior to the fabrication, a 200-nm-thick layer of SiO_2 was prepared by the thermal oxidation of a 1×1 cm^2 Si wafer. An EB resist mask was prepared on a 200-nm-SiO_2/Si substrate using an EB writer (Elionix, ESL-7500DEX, Tokyo, Japan). A 40-nmthick of Al with a 4-nm-adhesive Cr layer was then deposited on the substrate with the EB resist mask using EB deposition (UEP-300-2C, ULVAC, Kanagawa, Japan). After the lift-off step, an RIE process (ULVAC CE-300I, Kanagawa, Japan) with a mixture of CHF_3/O_2 (15-sccm/2.5-sccm) gases was applied to etch a 180-nmdepth SiO_2 layer with Al bowties as mask, forming a 180-nm height SiO_2 post underneath each Al triangle. The morphology of the fabricated Al bowtie arrays was characterized by using a field-emission scanning electron microscope (FE-SEM, Hitachi SU8000, Tokyo, Japan).

A con-focal Raman microscope (WITec Alpha 300S, Ulm, Germany) combined with a halogen lamp (Lucir-LAHL100) in UV-NIR range and a second harmonic diode pumped Nd:YAG laser (WITec, Ulm, Germany) at 532 nm was used for the investigation of the plasmonic scattering property (dark-field geometry, see Figure 1b) and the demonstration of the SERS effect (Figure 1c). As for the dark-field scattering measurements, a scanning mode equipped with a 50× dark-field lens (Zeiss, Oberkochen, Germany) was used. The relative scattering spectra were obtained by subtracting the spectra taken from the free-standing Al bowtie antennas with posts supported by the Si substrate to the spectra taken from a bare Si sample.

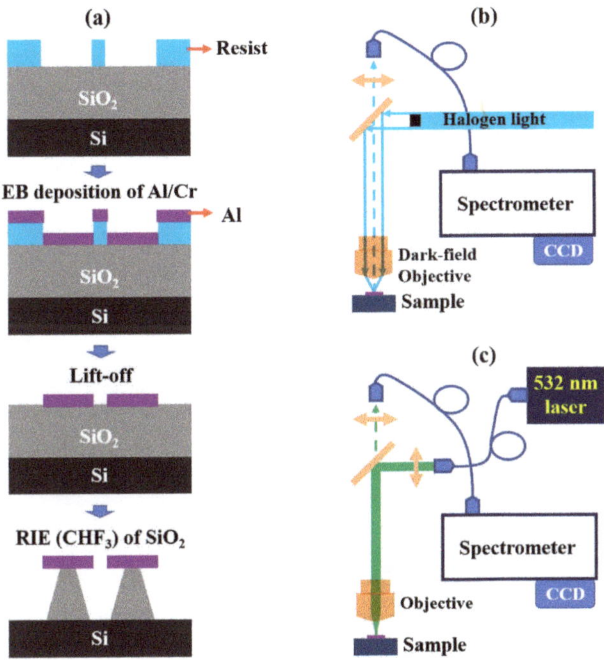

Figure 1. (**a**) Schematic illustration of the fabrication process. Optical microscopy setup of (**b**) dark-field measurement and (**c**) Raman measurement.

To demonstrate the SERS effect, a 10 μL droplet of Nile-blue solution (concentration 3 μM) was spread over the Al bowtie arrays (1 × 1 cm^2 substrate) and then dried in ambient condition. The SERS mapping measurements were performed on an area of 10 μm × 10 μm, with a spatial resolution of 40 pixel (250 nm/pixel), an integration time of 0.2 s/pixel. A 100× objective lens was used with an optimal laser power of 500 μW.

The rigorous coupled-wave analysis (RCWA) (DiffractMOD, Synopsys' RSoft, Version 2017.09, Ossining, NY, USA) and the finite-difference time-domain (FDTD) methods (FullWAVE, Synopsys' RSoft, Version 2017.09, Ossining, NY, USA) were employed to get further understanding on the optical properties of the plasmonic Al bowties as well as the enhancement nature of the electromagnetic field at their gaps and surfaces. The dielectric functions of Al, Cr, Si and SiO$_2$ were referred from the literature [55,56]. The model of the free-standing Al bowtie structures used for the simulations was constructed based on a build-in CAD layout (RSoft CAD Environment™, Version 2017.09, Ossining, NY, USA). Each free-standing bowtie antenna composed of a Si substrate and two SiO$_2$ posts (height of 180 nm) and Al bowties. The Al bowties have the triangle shapes (length of 105 nm, thickness of 40 nm), and composed of a 4-nmthick native Al$_2$O$_3$ oxide layer on top and 4-nm-thick Cr adhesive layer underneath. To mimic the real structures, the triangular truncations were placed at the corners of the triangles. The periodic boundary and grid size conditions were optimized to simulate the whole structure.

3. Results

Figure 2a presents the tilted-view FE-SEM image of an Al bowtie antenna array with parameters of 105 nm length, 20 nm gap, 1400 nm periodicity (the distance between two bowties), and 1120 nm spacing (the distance between two bowtie lines). The inset of Figure 2a shows a bowtie antenna standing freely on the SiO_2 posts. The spectroscopic measurements of the Al bowtie antennas are presented in Figure 2b,c. As the obtained spectra are achieved by subtracting the dark-field scattering spectrum taken from the area with the presence of bowtie antennas with respect to the reference area of the bare Si sample, it is hence inferred that our measurement reflects the plasmonic characteristics in the far-field of the antenna resonances. Under the near-field measurement, in the particular case of bowtie antennas, the resonance spectrum depends on the excited position of each single object. For example, the bowtie gap-mode (bright-mode) and dark-mode are seen under the excitation at the point near the gap, and at the point located far from the gap, respectively [57]. In the far-field measurement, the resonance spectrum is a superimposition of the scattering, emission, and interaction from the objects. There exists a difference between the near-field and far-field resonance. The far-field measurement reflects the resonant nature (the polarization is perpendicular or parallel to the bowtie axis) of the plasmonic nanoantennas and does not influence much on the SERS effect. In contrast, the near-field resonance does affect the SERS signals [58]. As seen in Figure 2b, the dark-field scattering spectral mapping of plasmonic resonance from the Al bowtie antenna array (taken in a range from 500 nm–700 nm of the resonance peak) indicates the far-field resonance of the antenna array, which reveals a clear spatial distribution of the circular spots following the periodicity of the actual antenna array.

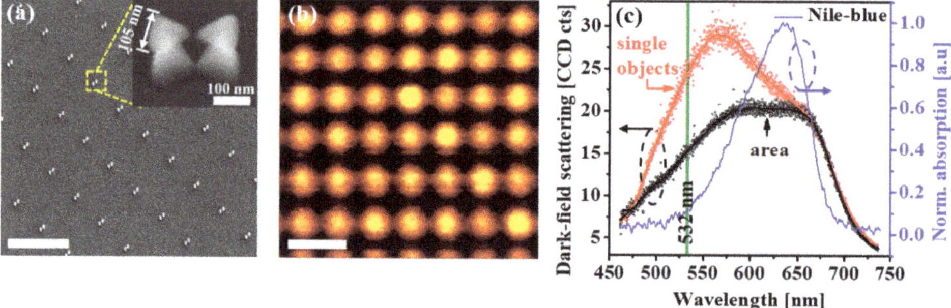

Figure 2. Spectral mapping of the free-standing Al bowtie antennas. (**a**) A scanning electron microscope (SEM) of the Al antenna arrays, observed at 30° oblique angle and the inset of one typical bowtie antenna. The antennas are spatially distributed with periodicity of 1400 nm, spacing of 1120 nm, 105 nm length and 20 nm gap. (**b**) The spectral mapping integrated over the resonance of the antennas, ranging from 500 nm to 700 nm. The circular "hot-spots" are spatially distributed following the SEM image, as shown in (**a**). The scale bar is 2 μm for both (**a**) and (**b**). (**c**) Spectral response of the single antenna (red), antenna array (black) and absorption spectrum of Nile-blue (blue), details in the text.

A comparison of the spectral characteristic between the single antenna and a scanned antenna array is shown in Figure 2c, together with the absorption spectrum of the organic Nile-blue solution, which was intentionally used as the probe molecules to demonstrate the field enhancement thanks to the SERS effect. The normalized dark-field scattering spectrum of the single objects shows a prominent plasmonic peak located at around ~565 nm and a noticeable hump at ~675 nm, toward the longer wavelength region. In the spectrum averaged from an antenna array consisting of the individual objects, the interaction between them and the surrounding areas, it is obvious that the low energy (long wavelength) peak still remained, but the main resonance peak is no more dominant.

Figure 3 shows the SERS measurements of the free-standing Al bowtie antennas with Nile-blue molecules. Very different from Au and Ag surfaces, which allow for achieving a well-defined chemical adsorption of a single layer of probe molecules (octadecanethiol (ODT), for instance), giving a precise estimation of the electromagnetic enhancement factor of the plasmonic nanoantennas [39]; the Al antennas are naturally coated by a few nanometer-thick Al_2O_3 that protects the Al surface from directly contacting the probe molecules. However, the Al_2O_3 is a fairly inert and reusable surface making it a promising SERS template for practical applications. The use of the Al_2O_3/Al configuration for the SERS measurements can solely demonstrate the electromagnetic field enhancement effect on SERS without any influence of the charge transfer (also known as chemical effect) between the plasmonic antennas and the Nile-blue molecules. The SERS performance of the free-standing Al bowtie arrays was investigated by evaluating the enhancement of the Raman spectrum of Nile-blue molecules. Figure 3a–c show an optical image (seen by confocal microscope), a reconstructed Raman mapping image (integrated at 590 cm^{-1} band), and a dark-field scattering mapping image of the corresponding area, respectively. Figure 3d presents the SERS spectra of Nile-blue molecules at the position "ON" the Al bowtie (red spectrum) and at the position "OFF" the Al bowtie (black spectrum). It is obvious that the Raman signals of Nile-blue fingerprints are strongly enhanced if the molecules are located at the position of Al bowties, giving clear evidence of the promising SERS effect due to the strong nearfield enhancement at the small gaps of the free-standing Al bowtie antennas. Additionally, the measurement of the polarization-dependent SERS was performed by using a half-wave plate to change the polarization of the excitation laser. The inset in Figure 3d shows that the relative intensity of Raman signals at the 590 cm^{-1} peak in the case of the excitation laser with polarization parallel to the bowtie axis (red curve) is significantly higher than that of the perpendicular polarization excitation (black curve). As the Nile-blue molecules cover the bowtie antennas entirely, the overwhelming Raman signal of parallel polarization compared to the perpendicular polarization excitation reflects that the gap-mode (excited under parallel polarization) gives a larger enhancement than the quadrupolar mode (excited under perpendicular polarization), which is in agreement with the previous report [59].

(a) Optical image   **(b)** Raman mapping   **(c)** Darkfield mapping

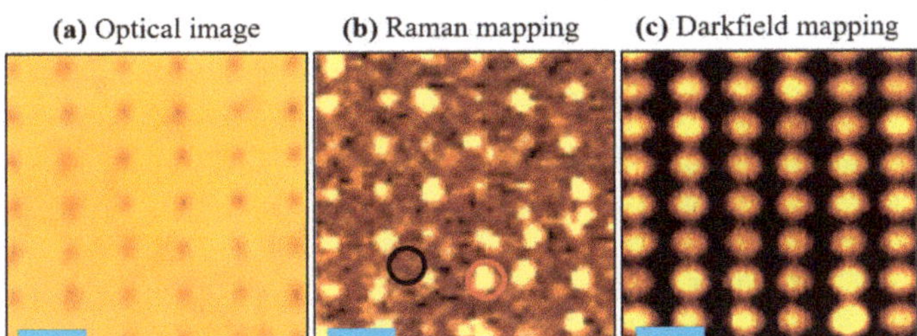

Figure 3. *Cont.*

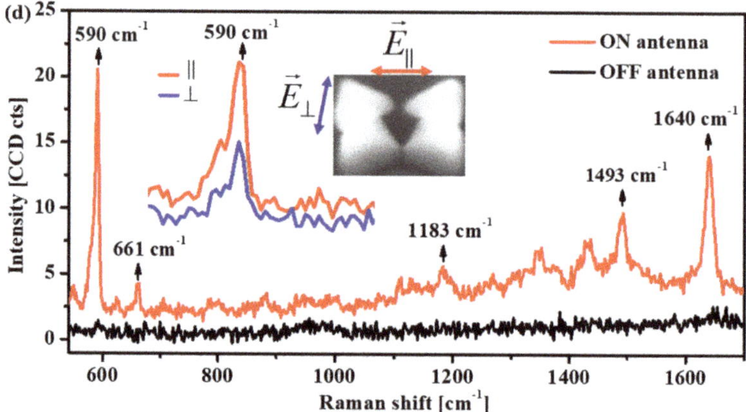

Figure 3. Demonstration of the electromagnetic field enhancement of free-standing bowtie antennas by SERS. (**a**) Optical image (seen under confocal microscope), (**b**) Raman mapping, and (**c**) corresponding plasmonic dark-field scattering mapping of the antenna arrays, respectively. The scale bar indicates 2 μm in all (**a**), (**b**), and (**c**). (**d**) SERS measurements of the Nile-blue molecules on the Al bowtie antennas. The black curve and red curve show Raman signals "OFF" and "ON" the antennas. The small inset indicates the Raman signals under different excitation conditions; with polarization parallel to the bowtie axis (red) and perpendicular to the bowtie axis (black), respectively.

The plasmonic resonance of the Al free-standing bowtie antennas was further elucidated using the numerical electromagnetic simulations; details are described in the Materials and Methods Section. Figure 4a shows the simulated extinction spectra of the floating Al bowtie antennas under two incident electric field polarizations; parallel to the bowtie axis with different gap sizes of 10 nm, 20 nm, 30 nm (solid curves), and perpendicular to the bowtie axis with agap of 10nm (dashed curve), separately. When the incident electric field oscillates parallel to the bowtie axis, the extinction spectrum for a 10-nm-gap antenna array shows a clear single resonance at 540 nm, and it is blue-shifted when the gap size increases. On the other hand, when the incident electric field oscillates perpendicular to the bowtie axis, the extinction spectrum of a 10-nm-gap bowtie antenna array tends to shift the resonance to the shorter wavelength region (see the dashed curve in Figure 4a). In comparison to the extinction spectrum measured on a single bowtie presented in Figure 2c, the simulated spectrum indicates a single peak, located closely to the main measured peak (560 nm). However, the simulation could not retrieve the second—(hump-like) peak at 670-nm. This discrepancy could be assigned to the fact that the free-standing architecture of the bowtie antennas results in a relatively strong far-field interaction, which could be seen only by far-field scattering experiments, not in the nearfield simulations [58]. Figure 4b shows the electric field distribution of Al bowtie antennas under the excitation of 532nm with the incident electric field polarized perpendicular to the bowtie axis. In this case, the excited hot-spots are located at the corners of each triangles and the field enhancement was found to be about E_x/E_{x0} = 7, where E_x and E_{x0} are the amplitudes of the enhanced and incident electric fields, respectively. Figure 4c,d show the electric field distribution of the Al bowtie antennas array with the incident electric field polarized parallel to the bowtie axis, under the same wavelength excitation of 532 nm; side-view (Figure 4c) and top-view (Figure 4d). It has been previously shown that the electric field enhancement decreases as the gap size of the bowtie increases [53,54]. As shown in Figure 4c,d, the electric field is mainly concentrated at the nanogap of the bowtie antennas. The enhanced area is homogeneously distributed over the gap and the enhancement factor (E_f) is calculated to be about $|E_x/E_{x0}| = 58$, which is much large rcompared to that of the perpendicular polarization excitation. It is consistent with the SERS results as discussed above where the SERS signal intensity under the electric field polarized parallel to the bowtie axis is stronger than that of the electric field polarized perpendicular to the bowtie axis (Figure 4d).

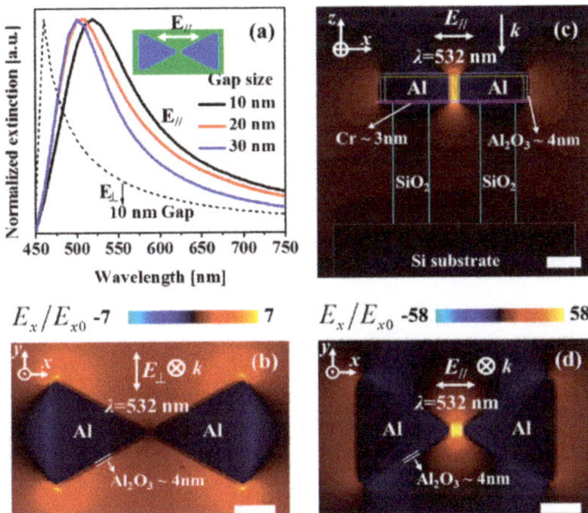

Figure 4. Simulated optical response and electric field distributions around Al bowtie excited at 532 nm wavelength. (**a**) Extinction spectra of the Al bowtie antennas at two different polarization conditions: parallel to the bowtie axis with different gap size of 10 nm, 20 nm, 30 nm (solid curves), and perpendicular to the bowtie axis with a 10 nm gap (dashed curve). The electric field distributions around a 105-nm-length, 10-nm-gap Al bowtie array excited at the 532-nm wavelength are simulated with two polarized excitations: (**b**) Perpendicular polarization, (**c**) and (**d**) parallel polarization with side-view and top-view, respectively. The scale bar is 50 nm for all panels (b), (c) and (d). The electric field hot-spot shows a strong electric field confinement with an enhancement factor $|E_x/E_{x0}|$ of 58, where E_x and E_{x0} are the amplitudes of the enhanced and incident electric fields, respectively.

4. Conclusions

In summary, we have numerically and experimentally demonstrated short-wavelength plasmonic Al bowtie antennas for SERS application at the 532-nm excitation. The free-standing Al bowtie antennas exhibited a strong electric field enhancement as high as 58 when excited at the 532 nm. We also further investigated optical properties of the fabricated free-standing Al bowtie antennas by introducing a plasmonic dark-field scattering mapping, which revealed the resonance of each individual antenna in the range between 500 nm–675 nm. Despite of having a natural oxide Al_2O_3 layer, and the absence of the chemical SERS effect, the correlated SERS mapping of Nile-blue molecules on Al bowties excited by a 532-nm laser clearly showed that the SERS signals were observed only at the bowtie antenna position. This evidences that the free-standing Al bowties exhibited a high SERS performance owing to the strongly confined electromagnetic field at the nanogaps between each triangle-pair of the bowtie antennas. Together with a recently proposed surface functionalization method of Al nanoantennas by using phosphoric acid [34], this work puts another step towards the practical applications of aluminum nano antennas in plasmonic sensing as well as in plasmon-enhanced nanospectroscopy.

Author Contributions: T.D.D. and C.V.H. performed the numerical simulations. A.O. and T.N. (Toshihide Nabatame) fabricated devices. C.V.H., T.D.D. conducted the spectroscopic measurements, characterized data. C.V.H., T.D.D. and T.N. (Tadaaki Nagao) co-wrote the manuscript with inputs from N.N. and N.Y. M.A. contributed to the discussions. T.N. (Tadaaki Nagao) supervised the project.

Acknowledgments: This work was partially supported by a Grant-in-Aid for Scientific Research (KAKENHI) projects (16F16315, JP16H06364, 16H03820) from the Japan Society for the Promotion of Science (JSPS), World Premier International Research Center Initiative on "Materials NanoArchitectonics" from MEXT (Japan) and CREST "Phase Interface Science for Highly Efficient Energy Utilization" (JPMJCR13C3) from Japan Science and

Technology Agency. C.V.H acknowledges the financial support from the Vietnam Academy of Science and Technology under the project VAST03.07/19-20.

Conflicts of Interest: The authors declare no conflict of interest.

References

1. Cortie, M.B.; Giddings, J.; Dowd, A. Optical properties and plasmon resonances of titanium nitride nanostructures. *Nanotechnology* **2010**, *21*, 115201. [CrossRef] [PubMed]
2. Naik, G.V.; Shalaev, V.M.; Boltasseva, A. Alternative Plasmonic Materials: Beyond Gold and Silver. *Adv. Mater.* **2013**, *25*, 3264–3294. [CrossRef] [PubMed]
3. Li, W.; Guler, U.; Kinsey, N.; Naik, G.V.; Boltasseva, A.; Guan, J.; Shalaev, V.M.; Kildishev, A.V. Refractory Plasmonics with Titanium Nitride: Broadband Metamaterial Absorber. *Adv. Mater.* **2014**, *26*, 7959–7965. [CrossRef] [PubMed]
4. Kumar, M.; Umezawa, N.; Ishii, S.; Nagao, T. Examining the Performance of Refractory Conductive Ceramics as Plasmonic Materials: A Theoretical Approach. *ACS Photonics* **2016**, *3*, 43–50. [CrossRef]
5. Dionne, J.A.; Sweatlock, L.A.; Sheldon, M.T.; Alivisatos, A.P.; Atwater, H.A. Silicon-Based Plasmonics for On-Chip Photonics. *IEEE J. Sel. Top. Quantum Electron.* **2010**, *16*, 295–306. [CrossRef]
6. Robusto, P.F.; Braunstein, R. Optical Measurements of the Surface Plasmon of Indium-Tin Oxide. *Phys. Status Solidi A* **1990**, *119*, 155–168. [CrossRef]
7. Ginn, J.C.; Jarecki, R.L.; Shaner, E.A.; Davids, P.S. Infrared plasmons on heavily-doped silicon. *J. Appl. Phys.* **2011**, *110*, 043110. [CrossRef]
8. Kanehara, M.; Koike, H.; Yoshinaga, T.; Teranishi, T. Indium Tin Oxide Nanoparticles with Compositionally Tunable Surface Plasmon Resonance Frequencies in the Near-IR Region. *J. Am. Chem. Soc.* **2009**, *131*, 17736–17737. [CrossRef]
9. Chen, K.; Guo, P.; Dao, T.D.; Li, S.; Ishiii, S.; Nagao, T.; Chang, R.P. Protein-Functionalized Indium-Tin Oxide Nanoantenna Arrays for Selective Infrared Biosensing. *Adv. Opt. Mater.* **2017**, *5*, 1700091. [CrossRef]
10. Jablan, M.; Buljan, H.; Soljačić, M. Plasmonics in graphene at infrared frequencies. *Phys. Rev. B* **2009**, *80*, 245435. [CrossRef]
11. Grigorenko, A.N.; Polini, M.; Novoselov, K.S. Graphene plasmonics. *Nat. Photonics* **2012**, *6*, 749. [CrossRef]
12. Ni, G.X.; Wang, L.; Goldflam, M.D.; Wagner, M.; Fei, Z.; McLeod, A.S.; Liu, M.K.; Keilmann, F.; Özyilmaz, B.; Castro Neto, A.H.; et al. Ultrafast optical switching of infrared plasmonpolaritons in high-mobility graphene. *Nat. Photonics* **2016**, *10*, 244. [CrossRef]
13. Lundeberg, M.B.; Gao, Y.; Asgari, R.; Tan, C.; Van Duppen, B.; Autore, M.; Alonso-González, P.; Woessner, A.; Watanabe, K.; Taniguchi, T.; et al. Tuning quantum nonlocal effects in graphene plasmonics. *Science* **2017**, *357*, 187–191. [CrossRef] [PubMed]
14. AlcarazIranzo, D.; Nanot, S.; Dias, E.J.C.; Epstein, I.; Peng, C.; Efetov, D.K.; Lundeberg, M.B.; Parret, R.; Osmond, J.; Hong, J.-Y.; et al. Probing the ultimate plasmon confinement limits with a van der Waals heterostructure. *Science* **2018**, *360*, 291–295. [CrossRef] [PubMed]
15. Ni, G.X.; McLeod, A.S.; Sun, Z.; Wang, L.; Xiong, L.; Post, K.W.; Sunku, S.S.; Jiang, B.-Y.; Hone, J.; Dean, C.R.; et al. Fundamental limits to graphene plasmonics. *Nature* **2018**, *557*, 530–533. [CrossRef] [PubMed]
16. Langhammer, C.; Schwind, M.; Kasemo, B.; Zorić, I. Localized Surface Plasmon Resonances in Aluminum Nanodisks. *Nano Lett.* **2008**, *8*, 1461–1471. [CrossRef]
17. Chan, G.H.; Zhao, J.; Schatz, G.C.; Van Duyne, R.P. Localized Surface Plasmon Resonance Spectroscopy of Triangular Aluminum Nanoparticles. *J. Phys. Chem. C* **2008**, *112*, 13958–13963. [CrossRef]
18. Knight, M.W.; Liu, L.; Wang, Y.; Brown, L.; Mukherjee, S.; King, N.S.; Everitt, H.O.; Nordlander, P.; Halas, N.J. Aluminum PlasmonicNanoantennas. *Nano Lett.* **2012**, *12*, 6000–6004. [CrossRef]
19. Sigle, D.O.; Perkins, E.; Baumberg, J.J.; Mahajan, S. Reproducible Deep-UV SERRS on Aluminum Nanovoids. *J. Phys. Chem. Lett.* **2013**, *4*, 1449–1452. [CrossRef]
20. West, P.R.; Ishii, S.; Naik, G.V.; Emani, N.K.; Shalaev, V.M.; Boltasseva, A. Searching for better plasmonic materials. *Laser Photonics Rev.* **2010**, *4*, 795–808. [CrossRef]
21. Jha, S.K.; Ahmed, Z.; Agio, M.; Ekinci, Y.; Löffler, J.F. Deep-UV Surface-Enhanced Resonance Raman Scattering of Adenine on Aluminum Nanoparticle Arrays. *J. Am. Chem. Soc.* **2012**, *134*, 1966–1969. [CrossRef]

22. Li, L.; Fang Lim, S.; Puretzky, A.A.; Riehn, R.; Hallen, H.D. Near-field enhanced ultraviolet resonance Raman spectroscopy using aluminum bow-tie nano-antenna. *Appl. Phys. Lett.* **2012**, *101*, 113116. [CrossRef]

23. Johnson, P.B.; Christy, R.W. Optical Constants of the Noble Metals. *Phys. Rev. B* **1972**, *6*, 4370–4379. [CrossRef]

24. Ehrenreich, H.; Philipp, H.R.; Segall, B. Optical Properties of Aluminum. *Phys. Rev.* **1963**, *132*, 1918–1928. [CrossRef]

25. Batson, P.E.; Silcox, J. Experimental energy-loss function, Im[$-1\varepsilon(q,\omega)$], for aluminum. *Phys. Rev. B* **1983**, *27*, 5224–5239. [CrossRef]

26. Lee, K.-H.; Chang, K.J. First-principles study of the optical properties and the dielectric response of Al. *Phys. Rev. B* **1994**, *49*, 2362–2367. [CrossRef]

27. Andersson, T.; Zhang, C.; Tchaplyguine, M.; Svensson, S.; Mårtensson, N.; Björneholm, O. The electronic structure of free aluminum clusters: Metallicity and plasmons. *J. Chem. Phys.* **2012**, *136*, 204504. [CrossRef] [PubMed]

28. Blaber, M.G.; Arnold, M.D.; Harris, N.; Ford, M.J.; Cortie, M.B. Plasmon absorption in nanospheres: A comparison of sodium, potassium, aluminium, silver and gold. *Phys. B Condens. Matter* **2007**, *394*, 184–187. [CrossRef]

29. Stöckli, T.; Bonard, J.-M.; Stadelmann, P.-A.; Châtelain, A. EELS investigation of plasmon excitations in aluminum nanospheres and carbon nanotubes. *Z. Phys. D Atoms Mol. Clust.* **1997**, *40*, 425–428.

30. Lehr, D.; Dietrich, K.; Helgert, C.; Käsebier, T.; Fuchs, H.-J.; Tünnermann, A.; Kley, E.-B. Plasmonic properties of aluminum nanorings generated by double patterning. *Opt. Lett.* **2012**, *37*, 157–159. [CrossRef]

31. Taguchi, A.; Saito, Y.; Watanabe, K.; Yijian, S.; Kawata, S. Tailoring plasmon resonances in the deep-ultraviolet by size-tunable fabrication of aluminum nanostructures. *Appl. Phys. Lett.* **2012**, *101*, 081110. [CrossRef]

32. Saito, Y.; Honda, M.; Watanabe, K.; Taguchi, A.; Song, Y.; Kawata, S. Design of Aluminum Nanostructures for DUV Plasmonics: Blue Shifts in Plasmon Resonance Wavelength by Height Control. *J. Jpn. Inst. Met.* **2013**, *77*, 27–31. [CrossRef]

33. Dao, T.D.; Chen, K.; Ishii, S.; Ohi, A.; Nabatame, T.; Kitajima, M.; Nagao, T. Infrared Perfect Absorbers Fabricated by Colloidal Mask Etching of $Al–Al_2O_3–Al$ Trilayers. *ACS Photonics* **2015**, *2*, 964–970. [CrossRef]

34. Chen, K.; Dao, T.D.; Ishii, S.; Aono, M.; Nagao, T. Infrared Aluminum Metamaterial Perfect Absorbers for Plasmon-Enhanced Infrared Spectroscopy. *Adv. Funct. Mater.* **2015**, *25*, 6637–6643. [CrossRef]

35. Fleischmann, M.; Hendra, P.J.; McQuillan, A.J. Raman spectra of pyridine adsorbed at a silver electrode. *Chem. Phys. Lett.* **1974**, *26*, 163–166. [CrossRef]

36. Jeanmaire, D.L.; Van Duyne, R.P. Surface Raman spectroelectrochemistry: Part I. Heterocyclic, aromatic, and aliphatic amines adsorbed on the anodized silver electrode. *J. Electroanal. Chem. Interfacial Electrochem.* **1977**, *84*, 1–20. [CrossRef]

37. Hartstein, A.; Kirtley, J.R.; Tsang, J.C. Enhancement of the Infrared Absorption from Molecular Monolayers with Thin Metal Overlayers. *Phys. Rev. Lett.* **1980**, *45*, 201–204. [CrossRef]

38. Nishikawa, Y.; Nagasawa, T.; Fujiwara, K.; Osawa, M. Silver island films for surface-enhanced infrared absorption spectroscopy: Effect of island morphology on the absorption enhancement. *Vib. Spectrosc.* **1993**, *6*, 43–53. [CrossRef]

39. Neubrech, F.; Pucci, A.; Cornelius, T.W.; Karim, S.; García-Etxarri, A.; Aizpurua, J. Resonant Plasmonic and Vibrational Coupling in a Tailored Nanoantenna for Infrared Detection. *Phys. Rev. Lett.* **2008**, *101*, 157403. [CrossRef]

40. Nie, S.; Emory, S.R. Probing Single Molecules and Single Nanoparticles by Surface-Enhanced Raman Scattering. *Science* **1997**, *275*, 1102–1106. [CrossRef]

41. Kneipp, K.; Kneipp, H.; Itzkan, I.; Dasari, R.R.; Feld, M.S. Surface-enhanced Raman scattering and biophysics. *J. Phys. Condens. Matter* **2002**, *14*, R597. [CrossRef]

42. Enders, D.; Nagao, T.; Nakayama, T.; Aono, M. In Situ Surface-Enhanced Infrared Absorption Spectroscopy for the Analysis of the Adsorption and Desorption Process of Au Nanoparticles on the SiO_2/Si Surface. *Langmuir* **2007**, *23*, 6119–6125. [CrossRef] [PubMed]

43. Nagao, T.; Han, G.; Hoang, C.; Wi, J.-S.; Pucci, A.; Weber, D.; Neubrech, F.; Silkin, V.M.; Enders, D.; Saito, O.; et al. Plasmons in nanoscale and atomic-scale systems. *Sci. Technol. Adv. Mater.* **2010**, *11*, 054506. [CrossRef] [PubMed]

44. Enders, D.; Nagao, T.; Pucci, A.; Nakayama, T.; Aono, M. Surface-enhanced ATR-IR spectroscopy with interface-grown plasmonic gold-island films near the percolation threshold. *Phys. Chem. Chem. Phys.* **2011**, *13*, 4935–4941. [CrossRef] [PubMed]

45. Hoang, C.V.; Oyama, M.; Saito, O.; Aono, M.; Nagao, T. Monitoring the Presence of Ionic Mercury in Environmental Water by Plasmon-Enhanced Infrared Spectroscopy. *Sci. Rep.* **2013**, *3*, 1175. [CrossRef] [PubMed]

46. Talley, C.E.; Jackson, J.B.; Oubre, C.; Grady, N.K.; Hollars, C.W.; Lane, S.M.; Huser, T.R.; Nordlander, P.; Halas, N.J. Surface-Enhanced Raman Scattering from Individual Au Nanoparticles and Nanoparticle Dimer Substrates. *Nano Lett.* **2005**, *5*, 1569–1574. [CrossRef] [PubMed]

47. von Maltzahn, G.; Centrone, A.; Park, J.-H.; Ramanathan, R.; Sailor, M.J.; Hatton, T.A.; Bhatia, S.N. SERS-Coded Gold Nanorods as a Multifunctional Platform for Densely Multiplexed Near-Infrared Imaging and Photothermal Heating. *Adv. Mater.* **2009**, *21*, 3175–3180. [CrossRef]

48. Zhang, C.-L.; Lv, K.-P.; Huang, H.-T.; Cong, H.-P.; Yu, S.-H. Co-assembly of Au nanorods with Ag nanowires within polymer nanofiber matrix for enhanced SERS property by electrospinning. *Nanoscale* **2012**, *4*, 5348–5355. [CrossRef]

49. Haynes, C.L.; Van Duyne, R.P. Plasmon-Sampled Surface-Enhanced Raman Excitation Spectroscopy. *J. Phys. Chem. B* **2003**, *107*, 7426–7433. [CrossRef]

50. Jiang, J.; Bosnick, K.; Maillard, M.; Brus, L. Single Molecule Raman Spectroscopy at the Junctions of Large Ag Nanocrystals. *J. Phys. Chem. B* **2003**, *107*, 9964–9972. [CrossRef]

51. Lim, D.-K.; Jeon, K.-S.; Kim, H.M.; Nam, J.-M.; Suh, Y.D. Nanogap-engineerable Raman-active nanodumbbells for single-molecule detection. *Nat. Mater.* **2009**, *9*, 60. [CrossRef] [PubMed]

52. Ye, J.; Wen, F.; Sobhani, H.; Lassiter, J.B.; Van Dorpe, P.; Nordlander, P.; Halas, N.J. Plasmonic Nanoclusters: Near Field Properties of the Fano Resonance Interrogated with SERS. *Nano Lett.* **2012**, *12*, 1660–1667. [CrossRef] [PubMed]

53. Schuck, P.J.; Fromm, D.P.; Sundaramurthy, A.; Kino, G.S.; Moerner, W.E. Improving the Mismatch between Light and Nanoscale Objects with Gold Bowtie Nanoantennas. *Phys. Rev. Lett.* **2005**, *94*, 017402. [CrossRef] [PubMed]

54. Hatab, N.A.; Hsueh, C.-H.; Gaddis, A.L.; Retterer, S.T.; Li, J.-H.; Eres, G.; Zhang, Z.; Gu, B. Free-Standing Optical Gold Bowtie Nanoantenna with Variable Gap Size for Enhanced Raman Spectroscopy. *Nano Lett.* **2010**, *10*, 4952–4955. [CrossRef] [PubMed]

55. Rakić, A.D.; Djurišić, A.B.; Elazar, J.M.; Majewski, M.L. Optical properties of metallic films for vertical-cavity optoelectronic devices. *Appl. Opt.* **1998**, *37*, 5271–5283. [CrossRef] [PubMed]

56. Palik, E.D. *Handbook of Optical Constants of Solids*, 3rd ed.; Academic Press: New York, NY, USA, 1998.

57. Fung, K.H.; Kumar, A.; Fang, N.X. Electron-photon scattering mediated by localized plasmons: A quantitative analysis by eigen-response theory. *Phys. Rev. B* **2014**, *89*, 045408. [CrossRef]

58. Doherty, M.D.; Murphy, A.; Pollard, R.J.; Dawson, P. Surface-Enhanced Raman Scattering from Metallic Nanostructures: Bridging the Gap between the Near-Field and Far-Field Responses. *Phys. Rev. X* **2013**, *3*, 011001. [CrossRef]

59. Gómez-Medina, R.; Yamamoto, N.; Nakano, M.; García de Abajo, F.J. Mapping plasmons in nanoantennas via cathodoluminescence. *New J. Phys.* **2008**, *10*, 105009. [CrossRef]

 *micromachines*

Article

MEMS-Based Wavelength-Selective Bolometers

Thang Duy Dao [1,*], Anh Tung Doan [1,2], Satoshi Ishii [1], Takahiro Yokoyama [1],
Handegård Sele Ørjan [1,2], Dang Hai Ngo [1,2], Tomoko Ohki [3], Akihiko Ohi [3], Yoshiki Wada [4],
Chisato Niikura [5], Shinsuke Miyajima [6], Toshihide Nabatame [3] and Tadaaki Nagao [1,2,*]

[1] International Center for Materials Nanoarchitectonics (MANA), National Institute for Materials
Science (NIMS), 1-1 Namiki, Tsukuba, Ibaraki 305-0044, Japan; doan.tunganh@nims.go.jp (A.T.D.);
sishii@nims.go.jp (S.I.); yokoyamatakahiro1007@gmail.com (T.Y.);
handegard.orjansele@nims.go.jp (H.S.Ø.); ngo.haidang@nims.go.jp (D.H.N.)
[2] Department of Condensed Matter Physics, Graduate School of Science, Hokkaido University,
Kita-10 Nishi-8, Kita-ku, Sapporo 060-0810, Japan
[3] Nanotechnology Innovation Station, National Institute for Materials Science (NIMS), 1-1 Namiki,
Tsukuba, Ibaraki 305-0044, Japan; ohki.tomoko@nims.go.jp (T.O.); ohi.akihiko@nims.go.jp (A.O.);
nabatame.toshihide@nims.go.jp (T.N.)
[4] Research Center for Functional Materials, National Institute for Materials Science (NIMS), 1-1 Namiki,
Tsukuba, Ibaraki 305-0044, Japan; wada.yoshiki@nims.go.jp
[5] Center for Green Research on Energy and Environmental Materials, National Institute for Materials
Science (NIMS), 1-1 Namiki, Tsukuba, Ibaraki 305-0044, Japan; niikura.chisato@nims.go.jp
[6] School of Engineering, Tokyo Institute of Technology, 2-12-1 Ookayama, Meguro-ku,
Tokyo 152-8550 Japan; miyajima.s.aa@m.titech.ac.jp
[*] Correspondence: dao.duythang@nims.go.jp (T.D.D.); nagao.tadaaki@nims.go.jp (T.N.);
Tel.: +81-29-860-4746 (T.N.)

Received: 13 May 2019; Accepted: 10 June 2019; Published: 21 June 2019

Abstract: We propose and experimentally demonstrate a compact design for membrane-supported wavelength-selective infrared (IR) bolometers. The proposed bolometer device is composed of wavelength-selective absorbers functioning as the efficient spectroscopic IR light-to-heat transducers that make the amorphous silicon (a-Si) bolometers respond at the desired resonance wavelengths. The proposed devices with specific resonances are first numerically simulated to obtain the optimal geometrical parameters and then experimentally realized. The fabricated devices exhibit a wide resonance tunability in the mid-wavelength IR atmospheric window by changing the size of the resonator of the devices. The measured spectral response of the fabricated device wholly follows the pre-designed resonance, which obviously evidences that the concept of the proposed wavelength-selective IR bolometers is realizable. The results obtained in this work provide a new solution for on-chip MEMS-based wavelength-selective a-Si bolometers for practical applications in IR spectroscopic devices.

Keywords: wavelength-selective sensors; infrared sensors; perfect absorbers; amorphous silicon; bolometers; microelectromechanical systems (MEMS)

1. Introduction

Microelectromechanical systems (MEMS, also known as micromachines) technology has been rapidly growing since the 1970s and early 1980s owing to its versatile application in a broad range of devices such as sensors, actuators, micropower generators and microfluidic systems [1–6]. Recently, MEMS have been identified as one of the most promising technologies for the industrial internet of things (IoT) market in the 21st century. Recent advances in photonics and nanofabrication techniques have proved the possibility in controlling optical properties of the photonic and metamaterial structures at the desired characteristics (i.e., transmittance, reflectance, absorptivity or polarization) and at the sub-wavelength partial scales. Merging photonic or metamaterial structures with MEMS has enabled a branch of micromachine system for optics and photonics which is so-called micro-opto-electro-mechanical systems (MOEMS) [7]. MOEMS have had a significant impact in optoelectronics ranging from lighting devices to imaging devices, especially in spectroscopic sensing devices such as wavelength selective infrared (IR) sensors and thermography [8,9].

A typical thermographic camera comprises an array of IR MEMS sensors that are designed in the mid-wavelength IR (MIR) atmospheric window (3 μm–5 μm or 9 μm–14 μm). The sensor is customarily made of a photoconductive layer such as InSb, InGaAs, HgCdTe or constructed as a quantum well IR photodetector (QWIP) [10,11]. Latterly, uncooled IR sensors such as thermopile [12,13] and bolometers (thermistors) including vanadium oxide (VO_x) [14–16] and amorphous silicon (a-Si) [17–20] have emerged as new alternative materials of choice for thermographic devices (microbolometers) because they do not require expensive cooling methods. Especially a-Si has shown great compatibility with the complementary metal–oxide–semiconductor (CMOS) and MEMS technologies [17,19,20].

The common pixel architecture of the microbolometer shows a resonant cavity constructed by a vertical standing wave Fabry–Pérot resonator that enhances the absorption of the incident IR radiation, which is typically designed in the range between 9 μm–14 μm. The resonance of the microbolometer can be further treated at the specific wavelengths to advantageously extend its applications to the multi-spectra IR chemical and medical imaging, IR remote sensing, and non-dispersive IR (NDIR) sensors. In this regard, multi-wavelength or tunable-wavelength microbolometers are desirable for the smart, multi-purpose and portable devices in practical applications. However, it is challenging to integrate multiple vertical-cavity resonators into a multi-wavelength sensors matrix. Another approach for spectroscopic microbolometers is to adopt lateral resonators such as plasmonic antennas [21–26], notably plasmonic perfect absorbers [27–32], which can energetically absorb and convert IR light into heat to induce the resonant response of bolometers. Furthermore, the lateral resonators in plasmonic perfect absorbers can be easily merged into a multi-wavelength sensor matrix.

The purpose of this work is to investigate the possibility to combine the lateral resonator perfect absorbers with simple MEMS-based bolometers for wavelength-selective IR sensors and multi-wavelength microbolometers. Here the proposed perfect absorber is constructed by a metal–insulator–metal configuration with the top metallic antennas array functioning as a plasmonic resonator [33–35]. This perfect absorber configuration has shown great potential in IR photonics because of its simple structure, wide working angle, polarization independence and wide resonance tunability. In particular, this perfect absorber can be easily arranged onto bolometers for wavelength-selective IR sensors. In this bolometer device, the efficient resonant IR energy absorbed in the absorber is directly converted into heat at the metals through Joule heating that follows Poynting's theorem; then conductively transfers to the bolometer which is thermally isolated from the surrounding media by a Si_3N_4 membrane. The reflectance spectra of the fabricated devices matched well with the pre-designed simulation performances. We experimentally showed that the resonance of the bolometer can be readily tuned just by changing the size of the plasmonic resonator. The measured spectral response of a typical device clearly showed the peak matching with the desired resonance. Therefore, the proposed wavelength-selective bolometers are feasible for the practical applications in multi-wavelength microbolometers and spectroscopic IR imaging devices.

The remainder of this paper is organized as follows. Section 2 describes the materials and methods used in this study. Section 3 presents the main results of this work specified in the following sub-sections. Section 3.1 discusses the details of the proposed membrane-supported wavelength-selective bolometers and its optical properties. Section 3.2 shows the fabrication of the a-Si and a-SiGe alloys for bolometers. The detailed fabrication process to realize the proposed bolometers is provided in Section 3.3. Then Section 3.4 shows the resonance tunability of the fabricated bolometers. As a proof of concept, the wavelength-selective responsivity of a typical fabricated bolometer is demonstrated in Section 3.5.

A short discussion on the potential applications of the proposed wavelength-selective bolometer is given in Section 3.6. Finally, Section 4 concludes the work.

2. Materials and Methods

Simulations: To determine the optimal geometrical parameters (i.e., size of square resonators, the periodicity, and the thickness of insulator- and metal films) for the proposed bolometers to have resonances at desirable wavelengths, numerical simulations were performed. Once the optimal geometries were obtained, we proceeded with fabrication to realize the device. Optical spectra including reflectance, transmittance and absorptivity of the proposed wavelength-selective bolometers were numerically simulated using the rigorous coupled-wave analysis (RCWA) method (DiffractMOD, Synopsys' RSoft). The electromagnetic field distributions and absorption maps of the proposed device were simulated using full-wave simulations based on the finite-difference time-domain (FDTD) method (FullWAVE, Synopsys' RSoft). The geometry of the device model was constructed to be identical to the design of the wavelength-selective bolometer using a computer-aided design (CAD) layout (RSoft CAD Environment™, Version 2017.09, Ossining, NY, USA). In the simulation, a 5-nm-thick Ti film serving as the adhesive layers was also added between each interface of Au and dielectric films. The dielectric functions of Au, Ti, Si, SiO_2 and Al_2O_3 were taken from literatures [36,37]. The dielectric function of Si_3N_4 was from the reference [38]. For the FDTD simulations, periodic boundary conditions were applied to both the x- and y-directions and perfectly matched layers were applied to the z-axis. In all the simulations, the incident electric field propagated along the z-axis and oscillated along the x-axis, the incident field amplitudes were normalized to unity.

Fabrications: The a-Si films were fabricated using a DC sputtering (sputter i-Miller CFS-4EP-LL, Shibaura, Yokohama, Japan) involving a boron-doped Si target (0.02 $\Omega \cdot$cm). For the a-SiGe alloys, a co-sputtering deposition was processed by adding a Ge target. The sputtering conditions of a-Si and a-SiGe alloys films were the same as follows; DC power of 100 W, Ar gas flow of 20 sccm, pressure of 0.304 Pa, sample holder rotation speed of 20 rpm and at room temperature. The membrane-supported wavelength-selective bolometers were processed on a 3-inch double side polished Si wafer. Prior to the fabrication of the devices, a 100-nm-thick SiO_2 layer was formed on both sides of the 3-inch Si wafer by the dry thermal oxidation at 1150 °C. Subsequently, a-350 nm thick Si_3N_4 film was sputtered on both sides of the SiO_2/Si wafer following a DC (200 W) reactive sputtering recipe from a boron-doped Si target in a mixture of Ar/N_2 (18/10 sccm) gases. The quality (hardness) of the Si_3N_4 films was then amended by a rapid thermal annealing (RTA) process in N_2 atmosphere (heating rate of 5 °C$\cdot$s^{-1}, keeping constant at 1000 °C for 1 min, then naturally cooling down). The fabrication of the MEMS devices used several steps of lithography processes including the direct laser writing (μPG 101 Heidelberg Instruments, Heidelberg, Germany) and electron beam writing (Elionix, ESL-7500DEX, Tokyo, Japan) combined with electron beam depositions of metals (UEP-300-2C, ULVAC, Yokohama, Japan) and reactive-ion etching (RIE) of Si_3N_4 (CHF$_3$ gas, Ulvac CE-300I). The elaborate fabrication of the membrane-supported wavelength-selective bolometers is detailed in the next section.

Characterizations: The amorphous structural property of fabricated a-Si films was verified using an X-ray diffractometer (XRD) with the Cu(Kα) line (SmartLab, Rigaku, Tokyo, Japan). The carrier characteristic of the fabricated bolometers was conducted using a Hall measurement system (Toyo Corporation, ResiTest 8400, Tokyo, Japan). The temperature-dependent resistance measurement was carried out using a source meter (Agilent Technologies B1500A, Santa Clara, CA, USA) combined with a temperature-controlled heating stage. The percent errors of all resistance measurements were less than 2%. The morphological characteristic of the fabricated membrane-supported bolometers was investigated using a scanning electron microscope (SEM) (Hitachi SU8230, Tokyo, Japan) operating at an accelerating voltage of 5 kV. The reflectance spectra of the fabricated devices were measured using an FTIR spectrometer (Nicolet iS50R FT-IR Thermo Scientific, Madison, WI, USA) equipped with a liquid N_2-cooled mercury cadmium telluride (MCT) detector and a KBr beam splitter. Then the absorptivity spectra were calculated by 1−reflectance, since the transmittance through the absorbers is zero with a 100-nm Au film serving as the bottom mirror. For the spectral response measurement, a tunable IR laser system (104 fs, 1 kHz repetition rate, Spectra-Physics) was used as a tunable excitation source [30].

3. Results and Discussions

3.1. Structural Design and Simulated Optical Properties

Figure 1a shows the proposed MEMS-supported wavelength-selective bolometer. The absorber layer placed on top of the bolometers composes of three layers; a top Au square antennas array serving as plasmonic resonators is isolated from a bottom Au planar mirror via an Al_2O_3 dielectric film. The absorbed IR energy in the absorber is converted to heat and then transfers to the a-Si bolometer film. The heat induces the change of the electrical resistance of the bolometer under an external bias applied on the two lateral Pt electrodes. The wavelength-selective absorber is thermally isolated to the Si substrate by a Si_3N_4 membrane.

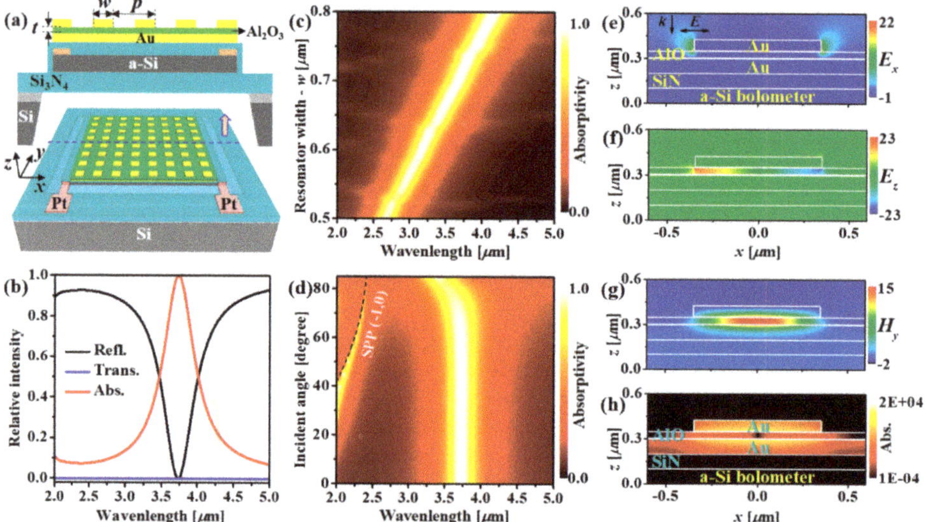

Figure 1. (**a**) Schematic illustrations, tilted view (bottom) and cross-sectional view (top), of the proposed MEMS-based wavelength-selective bolometer. (**b**) Simulated reflectance, transmittance and absorptivity of a device having geometrical parameters of p = 1.2 µm, t = 0.045 µm and w = 0.72 µm exhibits a resonance at 3.73 µm with a unity absorptivity. (**c**) Simulated absorptivity map of the chosen absorber for the proposed device show wide resonance tunability just simply by changing the size of the resonator while keeping other parameters unchanged (p = 1.2 µm, t = 0.045 µm). (**d**) Simulated angle-dependent absorptivity of the absorber reveals a wide-range working angle up to 70° of the proposed devices. The peak appeared in the shorter wavelength region indicates SPP in periodic Au square array. (**e**–**h**) Simulated electric field (E_x, E_z), magnetic field–H_y and absorption maps of a device excited at the resonance. In the simulations, the incident electric field propagated along the z-axis and oscillated along the x-axis. The thicknesses of the bottom Au layer and a-Si film were fixed at 0.1 µm and 0.2 µm for all simulations.

The geometrical parameters of the device including the width of Au square antenna—w, the periodicity of the square lattice—p, the thickness of the Al_2O_3 dielectric—t was optimized for certain desirable resonance. The thicknesses of the top Au antenna, the bottom Au film were fixed at 0.07 µm, 0.1 µm, respectively, which are much larger than the skin depth of Au (i.e., 0.03 µm) in the IR region. It is worth noting that the bottom Au film with a 0.1 µm thick is opaque in the MIR region, therefore the optical property of the absorber does not depend on layers underneath. Therefore, the resonance of the absorber relies mainly on p, t and w. Figure 1b plots the simulated reflectance (black curve), transmittance (blue curve) and absorptivity (red curve) of a device having the geometrical parameters of $p = 1.2$ μm, $t = 0.045$ μm and $w = 0.72$ μm. The device exhibits an evident resonance at 3.73 μm with almost unity absorptivity. If the periodicity and insulator layer are unchanged, the resonance wavelength of the device is directly proportional to the width (w) of the resonator while it retains perfect absorption in a wide spectral range of the MIR atmospheric window (Figure 1c). This provides a simple way to tune the active wavelength of the bolometer for spectroscopic applications such as for multi-wavelength detection. Further simulations were also performed to verify the effects of other geometrical parameters on optical spectra of the proposed device (Figure S1, Supplementary Information), including the resonator's thickness (Figure S1a), the insulator's thickness (Figure S1b), the periodicity (Figure S1c). In particular, with the symmetric resonator (square), the device's resonance does not depend on the polarization (Figure S1d). The proposed device can be also applied for the near IR and the long-wavelength IR regions (Figure S2, Supplementary Information). The angle-resolved absorptivity was also simulated, and the result is summarized in Figure 1d. Although there exists another resonance in the shorter wavelength region which is attributed to the surface plasmon polariton (SPP) at Au/air interface of the periodic Au square lattice array [38,39], the main resonance (magnetic mode) remains unchanged up to 70° incidence. This large working angle of the proposed device is essential for the practical applications.

To further understand the origin of the perfect resonant absorption in the device, we performed FDTD simulations to calculate electromagnetic field distributions of the absorber used in the proposed bolometer. Figure 1e–g shows the distributions of the electric fields (E_x and E_z) and magnetic field (H_y) of the device excited at the resonance (3.73 μm). In this metal–insulator–metal absorber, the top resonator functions as the sub-wavelength electric dipole antenna which defines the resonance of the absorber. The excited electric field of the electric dipole at the top resonator induces an inverse dipole at the bottom metal film. The oscillation of this antiparallel electric dipoles pair along with the excited electric field creates an electric current loop through the top resonator and bottom metal film, resulting in a large magnetic field enhancement between them (Figure 1g). The resonance of the absorber is so-called magnetic resonance. Interestingly, due to the optical loss of the metal, the absorbed energy at the resonance arises at the top Au resonator and bottom Au film subsequently converts to heat through Joule heating that follows Poynting's theorem [38,40]. The resonantly generated heat is then conductively transferred to the bolometer film via a thin Si_3N_4 layer to trigger the change of the electrical resistance for IR sensing. Here it is worthy that we adopted a 100-nm-thick Si_3N_4 layer to electrically isolate the metal film with the bolometer.

3.2. Amorphous Silicon and Silicon-Germanium Alloys for Bolometer

Prior to the fabrication of the membrane-supported bolometers, we fabricated bolometer films and examined their temperature coefficient of resistance (TCR) characteristics. In the past two decades, the a-Si and a-SiGe alloys bolometers (thermistors) have been intensively investigated in both scientific study and practical application due to its great success in achieving a high TCR with simple fabrication processes [20,41–43]. In this work, we also intended to choose a-Si film fabricated by sputtering for our MEMS-based wavelength-selective bolometers.

Figure 2a displays XRD patterns of two a-Si (p-type) films deposited on the 100-nm-thick SiO_2/Si substrates by DC sputtering using a boron-doped Si target; an as-deposited film shown in blue color and an annealed film for a comparison (500 °C in H_2 atmosphere) shown in red color. Both two films reveal the amorphous phase of the silicon with two broad 2θ peaks at 27.5° and 54° [44]. It is worth noting that the sharp peaks observed in the XRD patters are attributed to the Laue diffraction features of the Si substrate. The carrier characteristics of the as-deposited and annealed a-Si films were carried out via Hall measurements, which indicated that both films showed p-type carriers with concentrations of 7.23×10^{15} (cm^{-3}) and 7.08×10^{16} (cm^{-3}), respectively. The temperature-dependent resistance property of the a-Si films was performed to evaluate the TCR performance. In the semiconductor, the resistivity is the exponential function of thermal activation conductance expressed by [18]:

$$\rho = \rho_0 \exp\left(\frac{-E_a}{k_B T}\right) \tag{1}$$

where ρ, ρ_0 are the resistivities at a certain temperature and at the initial measured temperature, respectively. E_a, k_B and T are the activation energy, Boltzmann constant and temperature (K), respectively. Figure 2b presents the temperature-dependent resistivity of the as-deposited and annealed a-Si films. As seen, the resistivities of both the as-deposited and the annealed a-Si films decreases rapidly when the temperature increases. The resistance changes are plotted in $\ln(R/R_0)$, where R and R_0 are the resistances at temperature T and at the initial temperature, respectively, are inversely proportional to the increase of temperature (Figure 2c). The activation energies (E_a) of the two a-Si films were then obtained from the slope of $\ln(R/R_0)$ against the temperature changes, and the TCR values (α) were finally calculated using the following relation [18]:

$$\alpha = \frac{1}{\rho}\frac{d\rho}{dT} = \frac{E_a}{k_B T^2} \tag{2}$$

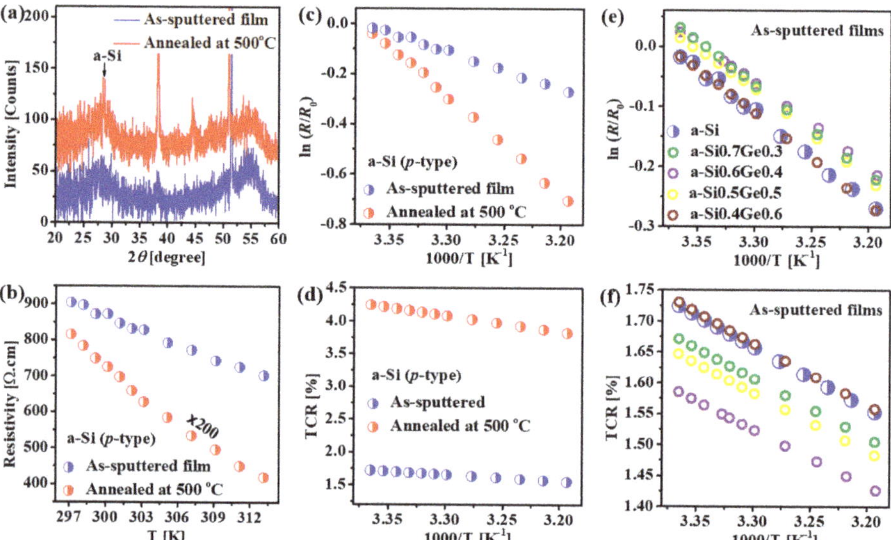

Figure 2. (**a**) XRD patterns and (**b**) Measured temperature-dependent resistivity curves of the as-sputtered (blue curve) and annealed at 500 °C (red curve) a-Si films. (**c**) Natural-log plots represented the change of resistance—$\ln(R/R_0)$ and (**d**) TCR values versus temperature changes of the as-sputtered (blue graphs) and annealed at 500 °C (red graphs) a-Si films. (**e**) Natural-log plots $\ln(R/R_0)$ and (**f**) TCR values versus temperature changes of a variety of amorphous Si-Ge alloys.

Figure 2d reveals the TCR values as a function of temperature. Although the TCR of the as-deposited a-Si film ~1.5% (K^{-1}) is lower compared to that of the 500 °C annealed a-Si film ~4% (K^{-1}), however, the value is rather good for bolometers. To be compatible with the fabrication of the elaborate MEMS-based devices, thermal annealing is not desirable. We also investigated a-Si$_{1-x}$Ge$_x$ alloys for bolometers with x variated from 0.3–0.7 (Figure 2e,f). As seen that the performance of the a-SiGe alloys are not much improved or even become worse compared to that of the pure a-Si film. Therefore, we chose the as-deposited a-Si for our MEMS-based wavelength-selective bolometers.

3.3. Fabrication of MEMS-Based Wavelength-Selective Bolometers

The MEMS-based wavelength-selective bolometers were then fabricated using several steps of lithography and following the structural parameters optimized by numerical simulations. The fabrication details are shown in Figure 3a–f. The devices were processed on a 3-inch double sides polished Si wafer with Si_3N_4/SiO_2 layered films on both sides (Figure 3a). An array of bolometers with different sizes (0.2×0.2 mm^2, 0.5×0.5 mm^2, 1×1 mm^2 and 2×2 mm^2) and with different wavelength resonances were arranged on the 3-inch wafer. Accordingly, the patterned a-Si films array and their lateral Pt electrodes were fabricated using two steps of lithography combined with sputtering deposition of a-Si and electron beam deposition of Pt (Figure 3b). Subsequently, a patterned 100-nm-thick Si_3N_4 layer was conformally deposited on each a-Si bolometer to electrically insulate the bolometer with the Au film of the absorber layer (Figure 3c). Then the wavelength-selective absorber on each bolometer were fabricated depending on the targeted resonance (Figure 3d,e). To process the membrane, a Si_3N_4 mask for the anisotropic wet-etching of Si on the back side of each bolometer was formed using RIE (CHF_3 gas) followed by a photoresist mask. Finally, the membrane-supported Si bolometers array was achieved by applying an anisotropic wet-etching at the back side of the Si wafer using a hot KOH solution (8 mg/l, 80 °C) (Figure 3f). Figure 3g–i explores the structural morphology of the fabricated MEMS-based wavelength-selective bolometers. Figure 3g,h shows typical top-view optical microscope images of the fabricated MEMS sensors, with two types of devices with 0.2×0.2 mm^2 and 2×2 mm^2 active areas. The inset in Figure 3g presents a photo of the MEMS sensors with light illuminated from the bottom which reveals the transparency of the Si_3N_4 membrane. Figure 3i presents a top-view SEM image of the fabricated absorber, which clearly shows the square resonator array of the absorber.

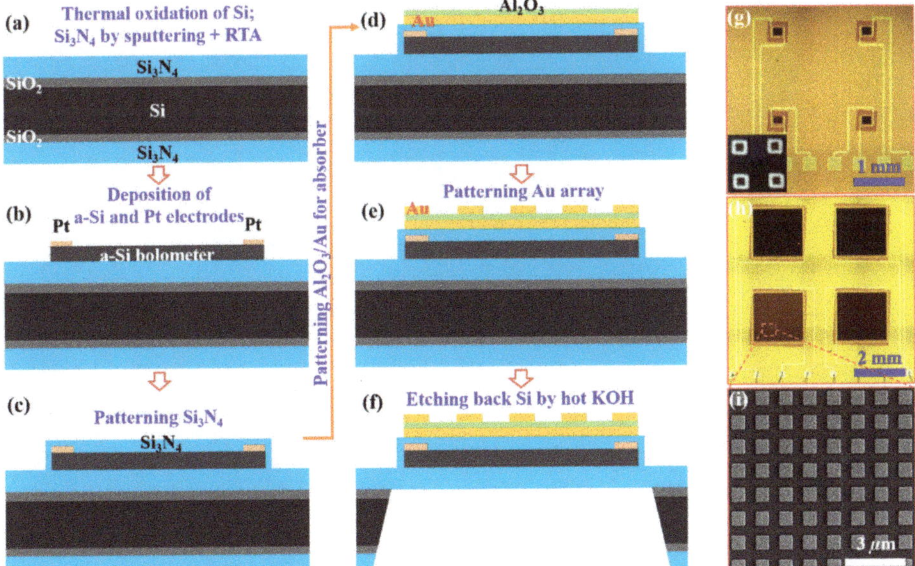

Figure 3. (**a–f**) Fabrication procedure of the MEMS-based wavelength-selective absorbers. (**g,h**) Top-view optical microscope images of the fabricated MEMS wavelength-selective bolometers with different square antenna sizes of the individual sensors. The inset in (**g**) reveals a photo of the MEMS-based quad-wavelength bolometer with the clear transparent Si_3N_4 membrane around each bolometer. (**i**) Top-view SEM image of the typical MEMS sensor having a resonance 3.65 µm.

3.4. Wavelength Tunability

The resonance spectra of the bolometers array were firstly characterized using an FTIR. As we discussed, the resonance of the bolometers can be readily tuned just by changing the size of the square resonator. Here we designed and fabricated a series of MEMS-based quad-wavelength bolometers chips. Figure 4a–d, from top to bottom, presents SEM images, simulated and measured absorptivities of the quad-wavelength bolometers. Four absorbers of the device have the same periodicity of 1.2 μm and the same insulator thickness of 0.045 μm but different resonators of 0.59 μm (Figure 4a), 0.64 μm (Figure 4b), 0.72 μm (Figure 4c) and 0.77 μm (Figure 4d) aiming at four different wavelengths of 3.11 μm 3.39 μm, 3.73 μm and 3.96 μm (middle panels of Figure 4). As seen in bottom panels of Figure 4, the fabricated MEMS-based quad-wavelength bolometers chip exhibits near-unity absorption resonances (>0.89) at four different wavelengths which are the same as the pre-designed resonances, evidencing the proper model and precise fabrication process used in this work. The resonances of the MEMS-based quad-wavelength bolometers presented here were chosen in the MIR atmospheric window; they can be also extended to the shorter or longer wavelength regions depending on the specific practical spectroscopic applications.

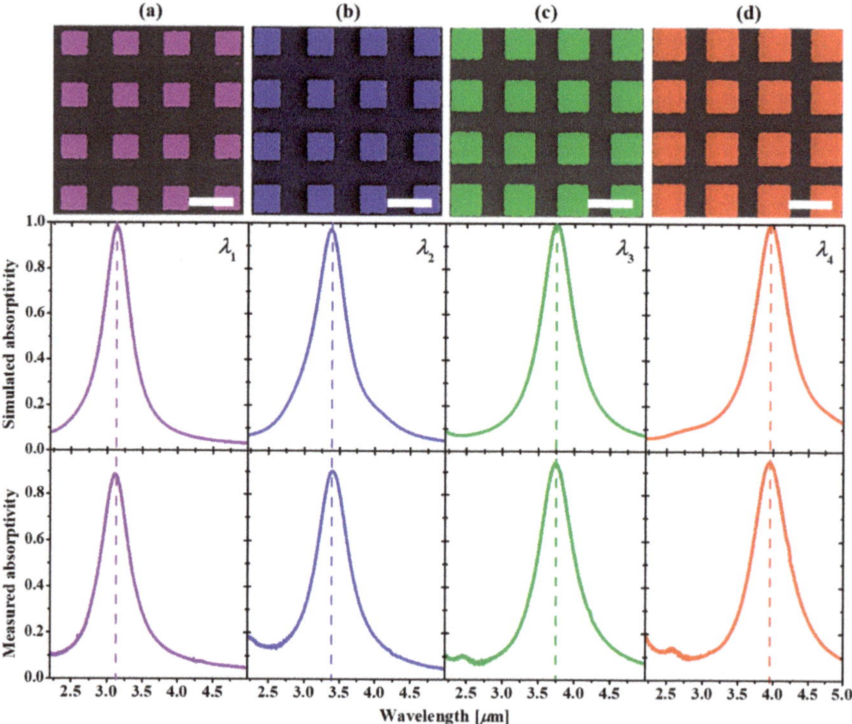

Figure 4. Resonance tunability in the MIR atmospheric window region of the MEMS wavelength-selective bolometers. (**a–d**) From top to bottom panels: top-view SEM images (with colors), simulated (middle) and measured (bottom) absorptivities of a quad-wavelength bolometers chip having resonances at 3.11 μm, 3.39 μm, 3.73 μm and 3.96 μm. The scale bar is 1 μm for all SEM images.

3.5. Wavelength-Selective Responsivity Measurement

To elucidate the spectral response characteristic of the proposed MEMS-based wavelength-selective bolometers, we performed the wavelength-dependent responsivity measurement. The measurement setup is illustrated in Figure 5a. In this measurement, a pulsed tunable IR laser (pulse width of 104 fs, repetition rate of 1 kHz) was used as the excitation source. The IR response of the sensor was measured via a read-out circuit, and then acquired using a source meter collected to a PC. The laser power at each wavelength was above the sensor's threshold. The spectral response of the sensor was finally normalized to the power spectrum of the laser.

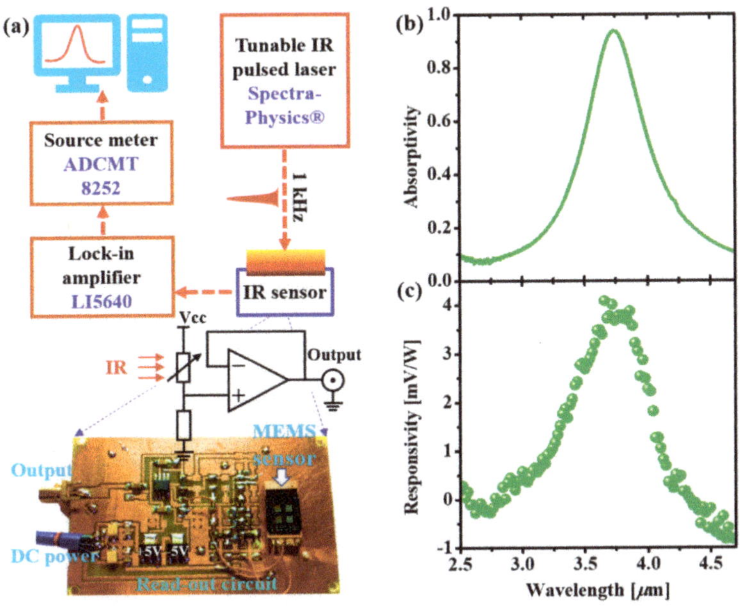

Figure 5. (a) Measurement setup of the spectral response of the fabricated bolometers. (b) Measured absorptivity and (c) Measured responsivity curves of a 3.73 μm resonant bolometer.

Figure 5b shows the measured absorptivity spectrum of a typical MEMS-based wavelength-selective bolometer having resonance at 3.73 μm. As we have discussed earlier, the resonantly absorbed IR light at the absorber is converted to heat, then conductively transfers to the bolometer to enable the change of the resistance. The resistance change can be measured as the IR response by applying a current through two Pt electrodes of the bolometer. Figure 5c displays the measured responsivity spectrum of the fabricated device. The measured spectrum responsivity curve exhibits a clear resonance. More interestingly, the resonance peak of the responsivity is the same as the absorptivity peak of the device (i.e., 3.73 μm), proving that the conceptual design for MEMS-based wavelength-selective bolometers is realizable. It should be noted that the measurement was done at a high frequency (1 kHz), because of which the responsivity was just a few mV/W. In the real applications, the device is expected to operate at the lower frequency such as a few hertz, such that the responsivity can be one to two orders higher compared to the value measured at the kilohertz frequency. In addition, the fabricated bolometer presented here have the size of a few hundred microns to millimeter scales; it can be scaled down to a few ten of microns, then the thermal isolation can be further improved for the higher responsivity and response time. The device is also compatible with the bridge-supported sensor technology which is accessible to the read-out integrated circuit (ROIC) for practical applications in IR imaging and thermography.

3.6. Potential Applications

Here we discuss some potential applications of the proposed MEMS-based wavelength-selective bolometer. As we mentioned earlier, the proposed device is devoted to serve as a new platform for IR spectroscopic sensors such as portable NDIR sensors, chemical IR imaging or IR remote sensing systems wherein the specific resonances are designed at vibrations of the targeted gases and molecules. The proposed wavelength-selective bolometer can be also applied for compact IR spectrometers, true temperature and emissivity measurements as well as multi-wavelength thermography by integrating multiple resonant sensors in a single chip.

4. Conclusions

We have successfully demonstrated a design for the MEMS-based wavelength-selective bolometers. The device used a-Si film deposited by DC sputtering at room temperature as the bolometer material. The fabricated a-Si film showed reasonably high TCR values above 1.5% (K^{-1}). A patterned metal–insulator–metal absorber structure was used as the efficient IR light-to-heat transducer for the wavelength-selective bolometers. The proposed bolometers revealed a facile resonance tunability just by changing the size of the patterned square metallic resonators arranged on top of each sensor. We also provided a detailed fabrication procedure to realize the proposed devices. As a proof of concept, we fabricated a set of MEMS-based quad-wavelength wavelength-selective bolometers having resonances in the MIR atmospheric window region. The fabricated devices exhibited the same resonances with the pre-designed devices. More interestingly, the measured spectral responsivity curve showed a clear wavelength-selective response following the absorptivity resonance, which proves the feasibility of the conceptual design of the MEMS-based wavelength-selective bolometers presented in this work. Although this work indicated resonances of bolometers in the MIR atmospheric window region, other resonant wavelengths and different number of resonances can be arranged in a single MEMS-based IR bolometers array chip depending on the specific practical applications such as NDIR sensors, IR spectroscopic sensing or imaging devices.

Supplementary Materials: The following are available online at http://www.mdpi.com/2072-666X/10/6/416/s1, Figure S1: Effects of the resonator thickness, insulator thickness, periodicity and polarization on optical spectra of the proposed wavelength selective bolometers; Figure S2: Simulated spectra of the proposed device with resonance in the near IR and long-wavelength IR regions.

Author Contributions: Conceptualization, T.N. (Tadaaki Nagao), T.D.D., S.I., A.O., T.N. (Toshihide Nabatame); device simulation, T.D.D., A.T.D., S.I.; bolometer fabrication, T.D.D., T.Y., C.N., S.M., A.T.D., D.H.N.; MEMS fabrication, T.O., T.D.D., A.O., T.Y.; device characterizations, T.D.D., A.T.D., H.S.Ø., D.H.N., S.I., Y.W.; data analysis, T.D.D., S.I., H.S.Ø.; writing—original draft preparation, T.D.D., T.N. (Tadaaki Nagao); writing—review and editing, all authors; supervision T.N. (Tadaaki Nagao); project administration, T.D.D. and T.N. (Tadaaki Nagao); funding acquisition, T.N. (Tadaaki Nagao).

Funding: This work was partially supported by JSPS KAKENHI (16F16315, JP16H06364, 16H03820), and CREST "Phase Interface Science for Highly Efficient Energy Utilization" (JPMJCR13C3) from Japan Science and Technology Agency.

Acknowledgments: The authors would like to thank all the staffs at Namiki Foundry, N. Furuhata, T. Yazawa and K. Okano at Photonics Nano-Engineering Group in NIMS for their kind technical supports. T. D. Dao would like to thank the fellowship program (P16315) from JSPS.

Conflicts of Interest: The authors declare no conflicts of interest.

References

1. Gabriel, K.; Jarvis, J.; Trimmer, W. *Small Machines, Large Opportunities: A Report on the Emerging Field of Microdynamics: Report of the Workshop on Microelectromechanical Systems Research*; Sponsored by the National Science Foundation; AT & T Bell Laboratories: Murray Hill, NJ, USA, 1988.
2. Bryzek, J. Impact of MEMS technology on society. *Sens. Actuators Phys.* **1996**, *56*, 1–9. [CrossRef]
3. Gardner, J.W.; Varadan, V.K. *Microsensors, Mems and Smart Devices*; John Wiley & Sons, Inc.: New York, NY, USA, 2001; ISBN 0-471-86109-X.
4. Judy, J.W. Microelectromechanical systems (MEMS): Fabrication, design and applications. *Smart Mater. Struct.* **2001**, *10*, 1115–1134. [CrossRef]
5. Bell, D.J.; Lu, T.J.; Fleck, N.A.; Spearing, S.M. MEMS actuators and sensors: Observations on their performance and selection for purpose. *J. Micromech. Microeng.* **2005**, *15*, S153–S164. [CrossRef]
6. Liu, C. *Foundations of MEMS*, 2nd ed.; Prentice Hall Press: Upper Saddle River, NJ, USA, 2011; ISBN 0-13-249736-0.

7. Motamedi, M.E. Micro-opto-electro-mechanical systems. *Opt. Eng.* **1994**, *33*, 3505–3517. [CrossRef]

8. Scribner, D.A.; Kruer, M.R.; Killiany, J.M. Infrared focal plane array technology. *Proc. IEEE* **1991**, *79*, 66–85. [CrossRef]

9. Cole, B.E.; Higashi, R.E.; Wood, R.A. Monolithic arrays of micromachined pixels for infrared applications. In Proceedings of the International Electron Devices Meeting 1998, Technical Digest (Cat. No.98CH36217), San Francisco, CA, USA, 6–9 December 1998; pp. 459–462.

10. Rogalski, A. Infrared detectors: An overview. *Infrared Phys. Technol.* **2002**, *43*, 187–210. [CrossRef]

11. Martyniuk, P.; Antoszewski, J.; Martyniuk, M.; Faraone, L.; Rogalski, A. New concepts in infrared photodetector designs. *Appl. Phys. Rev.* **2014**, *1*, 041102. [CrossRef]

12. Kanno, T.; Saga, M.; Matsumoto, S.; Uchida, M.; Tsukamoto, N.; Tanaka, A.; Itoh, S.; Nakazato, A.; Endoh, T.; Tohyama, S.; et al. Uncooled infrared focal plane array having 128×128 thermopile detector elements. *Proc. SPIE* **1994**, *2269*. [CrossRef]

13. Oliver, A.D.; Wise, K.D. A 1024-element bulk-micromachined thermopile infrared imaging array. *Sens. Actuators Phys.* **1999**, *73*, 222–231. [CrossRef]

14. Wood, R.A. Uncooled thermal imaging with monolithic silicon focal planes. *Proc. SPIE* **1993**, *2020*. [CrossRef]

15. Jerominek, H.; Picard, F.; Swart, N.R.; Renaud, M.; Levesque, M.; Lehoux, M.; Castonguay, J.S.; Pelletier, M.; Bilodeau, G.; Audet, D.; et al. Micromachined uncooled VO_2-based IR bolometer arrays. *Proc. SPIE* **1996**, *2746*. [CrossRef]

16. Wada, H.; Nagashima, M.; Kanzaki, M.; Sasaki, T.; Kawahara, A.; Tsuruta, Y.; Oda, N.; Matsumoto, S. Fabrication process for 256×256 bolometer-type uncooled infrared detector. *Proc. SPIE* **1997**, *3224*. [CrossRef]

17. Unewisse, M.H.; Liddiard, K.C.; Craig, B.I.; Passmore, S.J.; Watson, R.J.; Clarke, R.E.; Reinhold, O. Semiconductor film bolometer technology for uncooled IR sensors. *Proc. SPIE* **1995**, *2552*. [CrossRef]

18. Wood, R.A. Chapter 3 Monolithic Silicon Microbolometer Arrays. In *Semiconductors and Semimetals*; Kruse, P.W., Skatrud, D.D., Eds.; Elsevier: Amsterdam, The Netherlands, 1997; Volume 47, pp. 43–121. ISBN 0080-8784.

19. Tissot, J.-L.; Rothan, F.; Vedel, C.; Vilain, M. Jean-Jacques Yon LETI/LIR's Amorphous Silicon Uncooled Microbolometer Development. *Proc. SPIE* **1998**, *3379*. [CrossRef]

20. Vedel, C.; Martin, J.L.; Ouvrier-Buffet, J.L.; Tissot, J.L.; Vilain, M.; Yon, J.J. Amorphous-silicon-based uncooled microbolometer IRFPA. *Proc. SPIE* **1999**, *3698*. [CrossRef]

21. Chang, C.-C.; Sharma, Y.D.; Kim, Y.-S.; Bur, J.A.; Shenoi, R.V.; Krishna, S.; Huang, D.; Lin, S.-Y. A Surface Plasmon Enhanced Infrared Photodetector Based on InAs Quantum Dots. *Nano Lett.* **2010**, *10*, 1704–1709. [CrossRef] [PubMed]

22. Wu, W.; Bonakdar, A.; Mohseni, H. Plasmonic enhanced quantum well infrared photodetector with high detectivity. *Appl. Phys. Lett.* **2010**, *96*, 161107. [CrossRef]

23. Lee, S.J.; Ku, Z.; Barve, A.; Montoya, J.; Jang, W.-Y.; Brueck, S.R.J.; Sundaram, M.; Reisinger, A.; Krishna, S.; Noh, S.K. A monolithically integrated plasmonic infrared quantum dot camera. *Nat. Commun.* **2011**, *2*, 286. [CrossRef]

24. Ogawa, S.; Okada, K.; Fukushima, N.; Kimata, M. Wavelength selective uncooled infrared sensor by plasmonics. *Appl. Phys. Lett.* **2012**, *100*, 021111. [CrossRef]

25. Jing, Y.L.; Li, Z.F.; Li, Q.; Chen, X.S.; Chen, P.P.; Wang, H.; Li, M.Y.; Li, N.; Lu, W. Pixel-level plasmonic microcavity infrared photodetector. *Sci. Rep.* **2016**, *6*, 25849. [CrossRef]

26. Zheng, B.; Zhao, H.; Cerjan, B.; Yazdi, S.; Ringe, E.; Nordlander, P.; Halas, N.J. A room-temperature mid-infrared photodetector for on-chip molecular vibrational spectroscopy. *Appl. Phys. Lett.* **2018**, *113*, 101105. [CrossRef]

27. Maier, T.; Brückl, H. Wavelength-tunable microbolometers with metamaterial absorbers. *Opt. Lett.* **2009**, *34*, 3012–3014. [CrossRef] [PubMed]

28. Maier, T.; Brueckl, H. Multispectral microbolometers for the midinfrared. *Opt. Lett.* **2010**, *35*, 3766–3768. [CrossRef] [PubMed]

29. Niesler, F.B.P.; Gansel, J.K.; Fischbach, S.; Wegener, M. Metamaterial metal-based bolometers. *Appl. Phys. Lett.* **2012**, *100*, 203508. [CrossRef]

30. Dao, T.D.; Ishii, S.; Yokoyama, T.; Sawada, T.; Sugavaneshwar, R.P.; Chen, K.; Wada, Y.; Nabatame, T.; Nagao, T. Hole Array Perfect Absorbers for Spectrally Selective Mid-Wavelength Infrared Pyroelectric Detectors. *ACS Photonics* **2016**, *3*, 1271–1278. [CrossRef]

31. Suen, J.Y.; Fan, K.; Montoya, J.; Bingham, C.; Stenger, V.; Sriram, S.; Padilla, W.J. Multifunctional metamaterial pyroelectric infrared detectors. *Optica* **2017**, *4*, 276–279. [CrossRef]

32. Ahmed, A.; Kim, H.; Kim, J.; Hwang, K.; Kim, S. Enhancing the Responsivity of Uncooled Infrared Detectors Using Plasmonics for High-Performance Infrared Spectroscopy. *Sensors* **2017**, *17*, 908. [CrossRef] [PubMed]

33. Diem, M.; Koschny, T.; Soukoulis, C.M. Wide-angle perfect absorber/thermal emitter in the terahertz regime. *Phys. Rev. B* **2009**, *79*, 033101. [CrossRef]

34. Liu, X.; Starr, T.; Starr, A.F.; Padilla, W.J. Infrared Spatial and Frequency Selective Metamaterial with Near-Unity Absorbance. *Phys. Rev. Lett.* **2010**, *104*, 207403. [CrossRef]

35. Dao, T.D.; Chen, K.; Ishii, S.; Ohi, A.; Nabatame, T.; Kitajima, M.; Nagao, T. Infrared Perfect Absorbers Fabricated by Colloidal Mask Etching of Al–Al$_2$O$_3$–Al Trilayers. *ACS Photonics* **2015**, *2*, 964–970. [CrossRef]

36. Rakić, A.D.; Djurišić, A.B.; Elazar, J.M.; Majewski, M.L. Optical properties of metallic films for vertical-cavity optoelectronic devices. *Appl. Opt.* **1998**, *37*, 5271–5283. [CrossRef] [PubMed]

37. Palik, E.D. *Handbook of Optical Constants of Solids*, 3rd ed.; Academic Press: New York, NY, USA, 1998.

38. Dao, T.D.; Ishii, S.; Doan, A.T.; Wada, Y.; Ohi, A.; Nabatame, T.; Nagao, T. On-Chip Quad-Wavelength Pyroelectric Sensor for Spectroscopic Infrared Sensing. *Adv. Sci.* **2019**, in press. [CrossRef]

39. Dao, T.D.; Chen, K.; Nagao, T. Dual-band in situ molecular spectroscopy using single-sized Al-disk perfect absorbers. *Nanoscale* **2019**, *11*, 9508–9517. [CrossRef] [PubMed]

40. Poynting, J.H. XV. On the transfer of energy in the electromagnetic field. *Philos. Trans. R. Soc. Lond.* **1884**, *175*, 343–361.

41. García, M.; Ambrosio, R.; Torres, A.; Kosarev, A. IR bolometers based on amorphous silicon germanium alloys. *J. Non-Cryst. Solids* **2004**, *338–340*, 744–748. [CrossRef]

42. Syllaios, A.J.; Schimert, T.R.; Gooch, R.W.; McCardel, W.L.; Ritchey, B.A.; Tregilgas, J.H. Amorphous Silicon Microbolometer Technology. *MRS Proc.* **2000**, *609*, A14.4. [CrossRef]

43. Ambrosio, R.; Moreno, M.; Mireles, J., Jr.; Torres, A.; Kosarev, A.; Heredia, A. An overview of uncooled infrared sensors technology based on amorphous silicon and silicon germanium alloys. *Phys. Status Solidi C* **2010**, *7*, 1180–1183. [CrossRef]

44. Zeman, M.; van Elzakker, G.; Tichelaar, F.D.; Sutta, P. Structural properties of amorphous silicon prepared from hydrogen-diluted silane. *Philos. Mag.* **2009**, *89*, 2435–2448. [CrossRef]

Article

A MEMS-Based Quad-Wavelength Hybrid Plasmonic–Pyroelectric Infrared Detector

Anh Tung Doan [1,2,†], Takahiro Yokoyama [1,†], Thang Duy Dao [1], Satoshi Ishii [1], Akihiko Ohi [1], Toshihide Nabatame [1], Yoshiki Wada [3], Shigenao Maruyama [4] and Tadaaki Nagao [1,2,*]

[1] International Center for Materials Nanoarchitectonics, National Institute for Materials Science, 1-1 Namiki, Tsukuba 305-0044, Japan; doan.tunganh@nims.go.jp (A.T.D.); yokoyamatakahiro1007@gmail.com (T.Y.); dao.duythang@nims.go.jp (T.D.D.); sishii@nims.go.jp (S.I.); ohi.akihiko@nims.go.jp (A.O.); nabatame.toshihide@nims.go.jp (T.N.)

[2] Department of Condensed Matter Physics, Graduate School of Science, Hokkaido University, Kita-10 Nishi-8 Kita-ku, Sapporo 060-0810, Japan

[3] Research Center for Functional Materials, National Institute for Materials Science, 1-1 Namiki, Tsukuba 305-0044, Japan; wada.yoshiki@nims.go.jp

[4] National Institute of Technology, Hachinohe College, 16-1 Uwanotai, Tamonoki, Hachinohe City, Aomori Prefecture 039-1192, Japan; maruyama@ifs.tohoku.ac.jp

[*] Correspondence: nagao.tadaaki@nims.go.jp; Tel.: +81-29-860-4746

[†] These authors contributed equally to this work.

Received: 3 March 2019; Accepted: 29 March 2019; Published: 21 June 2019

Abstract: Spectrally selective detection is of crucial importance for diverse modern spectroscopic applications such as multi-wavelength pyrometry, non-dispersive infrared gas sensing, biomedical analysis, flame detection, and thermal imaging. This paper reports a quad-wavelength hybrid plasmonic–pyroelectric detector that exhibited spectrally selective infrared detection at four wavelengths—3.3, 3.7, 4.1, and 4.5 μm. The narrowband detection was achieved by coupling the incident infrared light to the resonant modes of the four different plasmonic perfect absorbers based on Al-disk-array placed on a Al_2O_3–Al bilayer. These absorbers were directly integrated on top of a zinc oxide thin film functioning as a pyroelectric transducer. The device was fabricated using micro-electromechanical system (MEMS) technology to optimize the spectral responsivity. The proposed detector operated at room temperature and exhibited a responsivity of approximately 100–140 mV/W with a full width at half maximum of about 0.9–1.2 μm. The wavelength tunability, high spectral resolution, compactness and robust MEMS-based platform of the hybrid device demonstrated a great advantage over conventional photodetectors with bandpass filters, and exhibited impressive possibilities for miniature multi-wavelength spectroscopic devices.

Keywords: infrared detector; quad-wavelength; hybrid plasmonic–pyroelectric; MEMS-based; spectral selectivity

---

1. Introduction

Multispectral selectivity is of crucial importance in the development of modern infrared (IR) detectors for modern spectroscopic applications including multi-wavelength pyrometry [1–5], non-dispersive infrared (NDIR) gas sensing [6–10], biomedical analysis [11–18], flame detection [19–22], and thermal imaging [23–25]. Spectrally selective IR detectors that are based on resonant cavity enhanced (RCE) photodetectors exhibit excellent spectral sensitivity and fast responses [26–31]. However, the requirement for cryogenic cooling makes them bulky, heavyweight, excessively costly, and complicated for some applications. Pyroelectric and thermopile detectors offer the advantages of being able to be operated at room temperature and of having wide spectral responses. Conventional spectrally selective uncooled detectors typically use passband filters mounted in front of the sensing element to filter out signals at the wavelengths that are out of interest, resulting in bulky designs and limited wavelength tunability.

Over the last two decades, the advent of plasmonic metamaterials, which are artificially structured materials with periodic subwavelength unit cells, has offered great freedom to tailor the absorption spectra [32–36]. The absorption peaks can be precisely controlled and manipulated by carefully designing the geometrical parameters of the unit cells. As the field of microelectromechanical systems (MEMS) has rapidly advanced, plasmonic perfect absorbers can be directly integrated on micromachined pyroelectric transducers to create compact, high-performance yet low-cost multi-wavelength detectors that operate at room temperatures.

In this work, we proposed and implemented a quad-wavelength pyroelectric detector with four distinct plasmonic absorbers to selectively detect light in the mid-IR region. For NDIR multi-gas sensing applications, the four resonance wavelengths were determined at 3.3, 3.7, 4.1, and 4.5 μm, which corresponded to the centered absorption band of CH_4, H_2S, CO_2, and N_2O [37,38]. The spectral selectivity was achieved by the coupling of incident infrared light to resonant modes of Al-disk-array/Al_2O_3/Al perfect absorbers with various disk sizes. The top patterned resonators were hexagonal arrays of disks used to achieve wide-angle acceptance and polarization-insensitivity, which are highly desirable for many sensing applications. We chose Al as the plasmonic base metal because it is abundant on Earth and it is industry-compatible while still exhibiting low-loss plasmonic properties similar to noble metals such as Au, Ag in the IR region [39]. The model of the Al-disk-array/Al_2O_3/Al perfect absorber was first constructed in a computer-aided design (CAD) layout (Rsoft CAD, Synopsys's Rsoft, Synopsys, Inc.) [40]. The absorptivities, electric field, and magnetic field distribution of the absorbers were simulated and optimized using the commercial rigorous coupled-wave analysis (RCWA) package and the FullWAVE package from Synopsys' Rsoft [40], which is a highly sophisticated tool for studying the interaction of light and photonic structures, including integrated wavelength-division multiplexing (WDM) devices [41,42], as well as nanophotonic devices such as metamaterial structures [34,43], and photonic crystals [44]. The sensing areas were designed as floating membranes above a void space to minimize thermal conduction, thereby improving the responsivity of the detector. The electromagnetic energy at the resonance wavelengths induced heat on the upper surface of the zinc oxide layer, which features pyroelectricity in thin film form. Due to the pyroelectric effect, a signal voltage was generated at the resonance wavelengths for each absorber. The on-chip design of the proposed quad-wavelength pyroelectric detector demonstrated the feasibility of integrating micro-detectors of different selective wavelengths into arrays with good CMOS compatibility. This opens the possibility of developing miniaturized and robust multi-color spectroscopic devices.

2. Design and Fabrication

2.1. Structure Design

The schematic diagram in Figure 1a illustrates the design layout of the proposed quad-wavelength detector. Four individual sensing elements were directly integrated on the same complementary metal-oxide-semiconductor (CMOS) platform with a size of 0.5×1.0 cm^2 to selectively detect IR radiation at four resonant wavelengths of 3.3, 3.7, 4.1, and 4.5 μm. The structural design of a single sensing element is illustrated in Figure 1b. From the top to bottom, it consisted of an Al-disk-array/Al_2O_3/Alperfect absorber structure with an active area of 200×200 μm^2, a 300 nm-thick pyroelectric zinc oxide thin film sandwiched between the Al back plate of the absorber and a 100 nm Pt/10 nm Ti bottom electrode, and a membrane-based CMOS substrate. A 300 nm-thick layer of silicon nitride was deposited on both sides of the silicon substrate to supply adequate mechanical strength for the membrane structure. The silicon wafer was 380 μm thick. The width and the length of the supporting arms were 20 μm and 400 μm. The cross-sectional profile of the unit cell is shown in Figure 1c. The Al-disk-array/Al_2O_3/Alperfect absorber consisted of an Al disk array as the resonator, an Al_2O_3 layer as the middle insulator, and an Al film as the back reflector. The thicknesses of the three layers were identical and were represented as t. The Al disks of diameter d were arranged in a periodic hexagonal array of periodicity p. After some numerically computational efforts based on the RCWA analysis, the optimized geometrical dimensions for the perfect absorbers were taken as p = 2.0 µm, t = 100 nm, and d = 0.97, 1.25, 1.35, 1.60 µm for the individual sensing elements at resonance wavelengths of 3.3, 3.7, 4.1, and 4.5 µm, respectively. When an incident IR radiation impinged on the top surface of the device, its external oscillating electric field induced electric dipoles inside the Al disks, which excited anti-parallel currents between the disk array and the back reflector. These circular currents produced a magnetic flux opposing the external magnetic field, resulting in a magnetic resonance. By adjusting the geometrical parameters of the Al disk-array, we could tailor the electromagnetic response of the structure to the external electromagnetic field to achieve selective absorption.

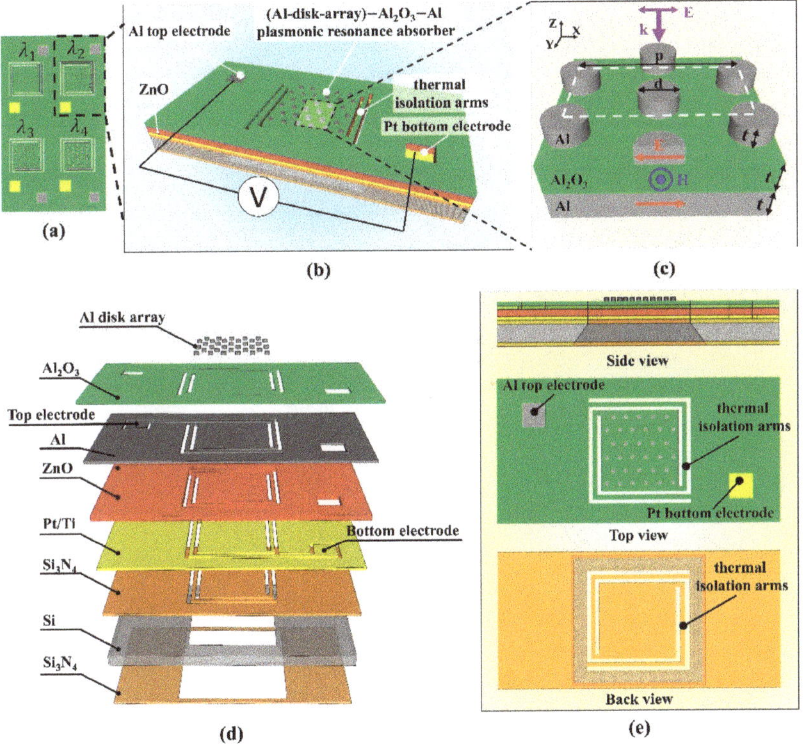

Figure 1. (**a**) Schematic illustration of the proposed quad-wavelength detector. (**b**) Illustration of the structural design of a single sensing element. (**c**) Illustration of the plasmonic perfect absorber with the indicated field lines at resonance (electric field E, red arrows; magnetic field H, blue arrows). (**d**) Exploded view of a single sensing element. (**e**) Side view, top view, and bottom view of a single sensing element.

The exploded view of a single sensing element is shown in Figure 1d. It was shown that the bottom layer of plasmonic metamaterial perfect absorber was also utilized as the top electrode of the pyroelectric detector. In this design, the zinc oxide layer was stacked below and in intimate contact with the plasmonic resonance absorber to fully exploit the spectrally selective radiation. It is also worth noting that we chose zinc oxide as the pyroelectric material due to its advantages of being nontoxic and compatible with the semiconductor process. This design is also applicable for other pyroelectric materials such as lead zirconate titanate, lithium tantalate, lithium niobate, barium strontium titanate, deuterated triglycine sulfate, and others. The side view, top view, and back view of the single sensing elements are shown in Figure 1e, indicating the micromachined floating membrane design of the sensing element.

2.2. Simulation

To tailor and optimize the spectrally selective absorption, the optical properties of the plasmonic absorbers were simulated by implementing rigorous coupled-wave analysis (RCWA) (DiffractMOD, Synopsys' Rsoft). Figure 2a shows the simulated reflectivity, transmissivity, and absorptivity characteristics of the two Al-disk-array/Al$_2$O$_3$/Al perfect absorbers which had diameters of 1.25 and 1.6 µm, respectively. As the optimal parameters, the unit cells had periodic dimensions of 2.0 µm, and the thicknesses of the Al disks, Al$_2$O$_3$ dielectric spacer and Al back reflector were all 100 nm. Since the continuous metal back reflector was thick enough to block light transmission, the transmissivity was zero across the investigated wavelength range. With disk diameters of 1.25 and 1.6 µm, the simulated absorptivities at normal incidence were resonantly enhanced up to 0.99 and 0.94 at the resonance wavelengths of 3.72 and 4.05 µm, with a full width at half maximum (FWHM) of 0.44 and 0.63 µm, respectively.

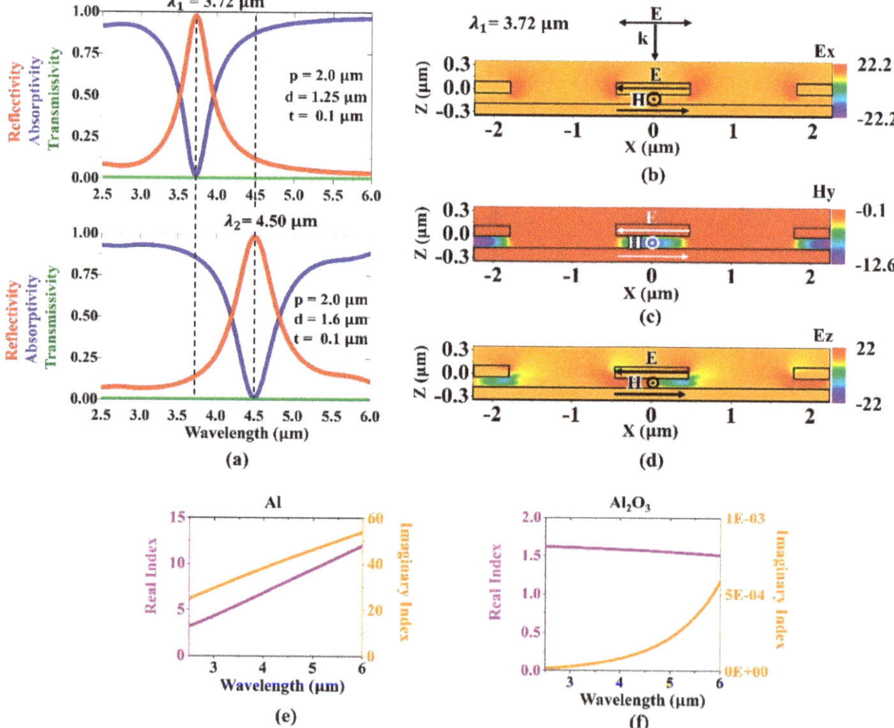

Figure 2. (a) Reflectivity, absorptivity, and transmissivity characteristics of the plasmonic perfect absorber at the resonance wavelengths of 3.7 µm and 4.50 µm. The simulated amplitude plots of (b) electric field in x direction, (c) magnetic field in y direction, (d) electric field in z direction at the resonance wavelength of 3.7 µm, illustrates the highly localized characteristics of the plasmonic perfect absorber. The refractive index dispersion data of (e) Al and (f) Al$_2$O$_3$ used in the simulations was taken from literature [45].

To gain insight into the resonant behavior of the Al-disk-array/Al$_2$O$_3$/Al perfect absorber, a full wave simulation based on the finite-difference time-domain (FDTD) method (FullWAVE, Synopsys' Rsoft) was performed at the resonance wavelength of 3.72 μm. The dielectric functions of Al, Al$_2$O$_3$ were taken from the literature. The simulated electric field distribution in x direction (Figure 2b) shows that the electric field was localized at the edges of the Al disks, indicating the induced electric dipoles. The magnetic field was intensively confined under the center of the top resonator (Figure 2c), for which the antiparallel currents were excited in the top Al disks and the bottom Al reflector (Figure 2d).

The dependence of the resonant modes of the plasmonic absorber as a function of disk size is illustrated in Figure 3a. Accordingly, the peak absorption was found to shift almost linearly with the disk diameter. Meanwhile, the resonance wavelength was not greatly affected by the variations in periodicity of the unit cells (Figure 3b). Due to this robust tunability with respect to the disk size, the electromagnetic response of these Al-disk-array/Al$_2$O$_3$/Al structures were engineered to achieve resonant absorption at multiple desired wavelengths.

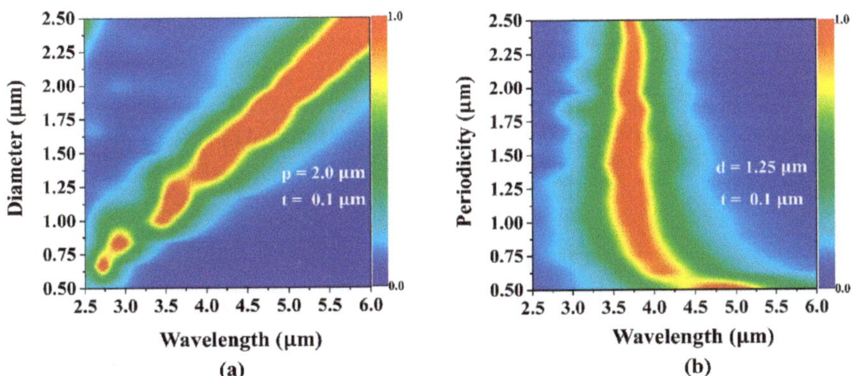

Figure 3. (a) The periodicity dependence with p changed from 0.5 to 2.5 μm while keeping the diameter and insulator thickness unchanged at 1.25 μm and 0.1 μm, respectively. (b) The diameter dependence with d changed from 0.5 to 2.5 μm, while keeping the periodicity and insulator thickness unchanged at 2.0 μm and 0.1 μm, respectively.

Figure 4a shows the simulated incident angle-dependent absorptivity with the geometrical parameters p = 2.0 μm, d = 1.25 μm, and t = 0.1 μm. The simulation results showed that the proposed plasmonic perfect absorber exhibited strong absorption independence to the incident angles. The absorption still remained near unity when the incident angle was increased up to 85°. Figure 4b shows the polarization-dependent absorptivity with the geometrical parameters p = 2.0 μm, d = 1.25 μm, and t = 0.1 μm. The result indicated that the resonances of the plasmonic absorber were almost unchanged when the polarization angle changed from 0 to 90°. The symmetrical design typically resulted in a wide-angle acceptance and polarization independence, which was desirable for most of the sensing applications.

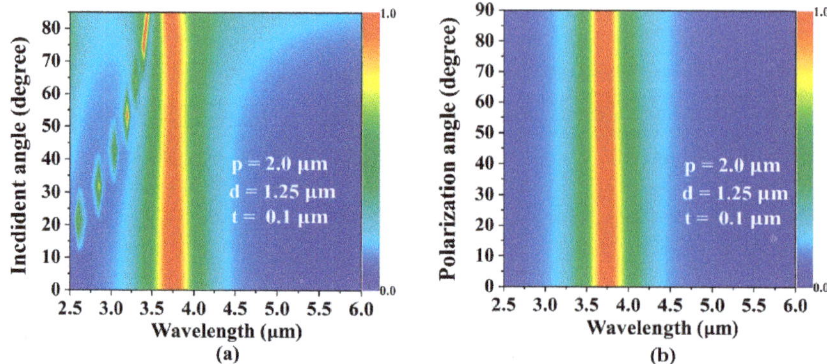

Figure 4. Simulated absorptivity spectrum as a function of (**a**) the incident angle and (**b**) polarization angle for an absorber with p = 2.0 µm, d = 1.25 µm, and t = 0.1 µm.

2.3. Fabrication

The fabrication of the proposed detector involved standard photolithography with dry and wet etching techniques. The microfabrication process of the MEMS-based hybrid plasmonic–pyroelectric detector is shown in Figure 5. The device was fabricated on commercial double-side polished *n*-type silicon prime wafers with <100> orientation in cleanroom facilities.

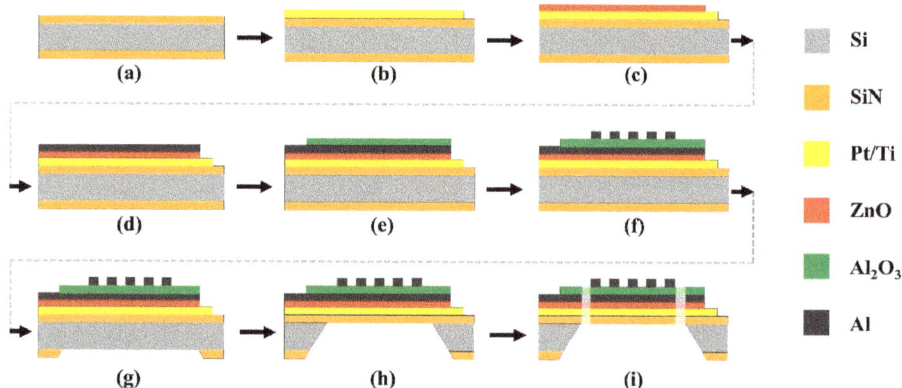

Figure 5. CMOS-compatible microfabrication process: (**a**) Rapid thermal oxidation of the silicon wafer and deposition of Si3N4 on both sides; (**b**) Deposition and lift-off of the Pt bottom electrode; (**c**) Deposition and lift-off of ZnO; (**d**) Deposition and lift-off of Al; (**e**) Deposition and lift-off of Al_2O_3; (**f**) E-beam lithography of the Al disk; (**g**) Dry etching backside of Si_3N_4; (**h**) Wet etching of Si by KOH; (**i**) Dry etching of Si_3N_4.

First, an 80 nm-thick SiO_2 insulator was formed by the dry thermal oxidation of the silicon wafer. The fabrication process was followed by depositing a 300 nm-thick silicon nitride (Si_3N_4) layer onto both sides of the silicon substrate by radio frequency (RF) reactive sputtering using a boron-doped p-type Si target processed in ambient N_2 as a reactive gas with a flow rate of 20 sccm (sputter i-Miller CFS-4EP-LL, Shibaura, Tokyo, Japan) (Figure 5a). This silicon nitride layer isolated the thermal conduction to the substrate, which could enhance the pyroelectric signal and the overall responsivity of the device. Then the silicon nitride layer was put into a rapid thermal annealing process at 1000 °C for 1 min. After that, laser direct writing lithography (µPG 101 Heidelberg Instruments) was performed to pattern the bottom electrode on top of the silicon nitride layer with a double layer of resists. The

100 nm Pt/10 nm Ti bottom electrodes were deposited in an electron beam evaporator at ambient temperatures, followed by a lift-off process in acetone (Figure 5b). Then, the sputter deposition of the 300 nm-thick zinc oxide layer was followed by laser direct writing lithography with positive photoresist combined with reactive-ion etching (RIE) using CF_4 gases (Ulvac CE-300I) (Figure 5c). Next, the sputter deposition of the 100-nm-thick Al top electrode was followed by laser direct writing lithography with positive photoresist, and RIE using BCl_3/Cl_2 gases (Figure 5d). After exposing and developing the pattern, a 100 nm Al_2O_3 film was sputtered and then patterned by direct laser writing lithography combined with a RIE step using CF_4 gas (Figure 5e). The 100-nm-thick Al disk array was patterned using the e-beam lithography and lift-off process (Figure 5f). The Si_3N_4 back layer serving as a RIE mask for the wet etching process of Si was patterned by RIE with CHF_3 gas using a photoresist mask formed by direct laser writing lithography (Figure 5g). Then, the backside wet alkaline etching was implemented as follows (Figure 5h). A thin polymeric ProTEK®B3 protective coating was spun on the front side of the structure to provide a temporary wet-etch protection for the patterned Al-disk-array/Al_2O_3/Al resonator during alkaline etches. The Si layer from the back side of each sensor was completely released by anisotropic wet-etching by immersing the structure in a deep reactive KOH 40% aqueous solution, heating at 60 °C for 10 h. Then, the polymeric protective coating was removed by immersing it into acetone. Finally, reactive-ion etching of Si_3N_4 was carried out to form the supporting arms (Figure 5i). The Si_3N_4/SiO_2 composite layer was used as the membrane to support the zinc oxide pyroelectric film and absorber. Each sensing element was wire-bonded to the Cu electrodes attached on a glass plate. The materials and the fabrication process of the detector was fully compatible with the CMOS.

3. Results and Discussion

Figure 6a shows a photo of a fabricated MEMS-based hybrid plasmonic–pyroelectric detector, which had a width of 0.5 cm and length of 1 cm. It clearly demonstrates that multiple hybrid plasmonic–pyroelectric sensing elements were easily integrated on a standard CMOS platform to achieve multispectral selectivity without any additional bulky optical filters. Figure 6b shows the optical microscopy images of the top view and back view of a single sensing element with an active area of 200 × 200 µm. It verifies that the sensing area was suspended by the long thermal isolation arms. The suspended area remained flat without any additional stress-reducing process.

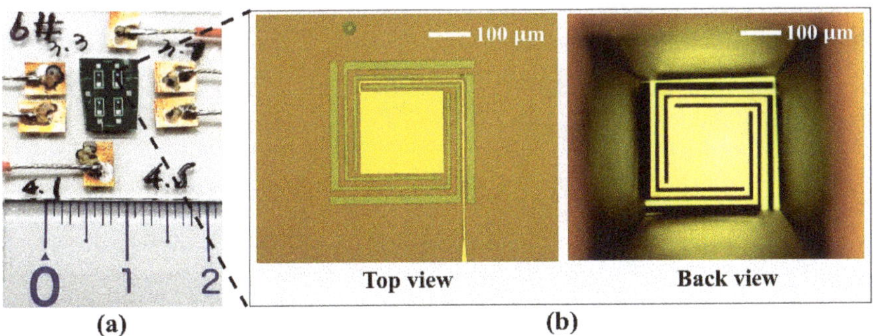

Figure 6. (a) Photo of a fabricated MEMS-based hybrid plasmonic–pyroelectric detector. (b) Microscope images of the top view and back view of the fabricated sensing element.

While the thicknesses of the sputtered layers could be precisely controlled by establishing optimum sputtering conditions and deposition rates, the process of transferring the shape and size of the Al disk arrays pose further challenges due the complex multi-stage photolithography of the sub-micron patterning. The residual photoresist and non-uniformity may result in degradation from the expected performance. Figure 7a–d show the scanning electron microscope (SEM) images of the hexagonal arrays of the Al disk resonators with diameters of 0.97, 1.25, 1.35, 1.60 µm. It was shown that the Al disk arrays fabricated on top of the membrane structures were well-defined and homogeneously distributed, indicating that the patterning process was precisely implemented. Because the Al disk arrays were patterned using electron-beam lithography with sub-10 nm resolution, there was a tolerance of a few nm in the diameter of the disks.

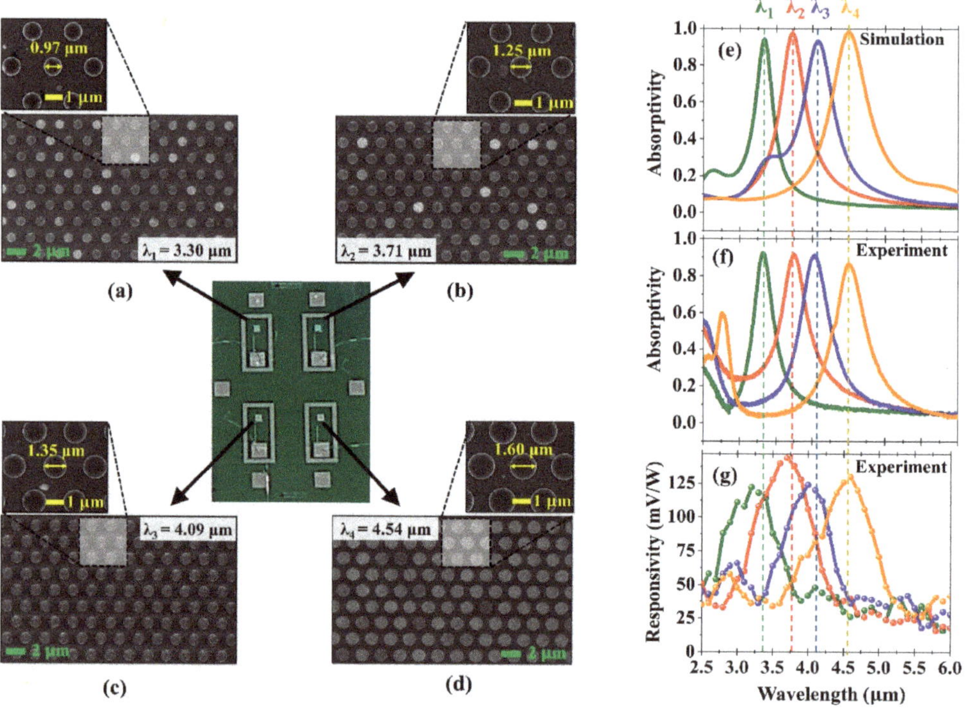

Figure 7. (**a–d**) Scanning electron microscope (SEM) images of the fabricated disk array patterns at four resonance wavelengths. (**e,f**) Simulated and measured absorptivities of the perfect absorbers at four resonance wavelengths. (**g**) Measured responsivity of the quad-wavelength detector.

The reflectance spectrum of each sensing element was measured using a Fourier transform infrared spectrometer (FTIR) (Thermo Scientific Nicolet iS50, Thermo Fisher Scientific, Waltham, MA, USA), coupled with a microscope (Nicolet Continuum FTIR Microscope, Thermo Nicolet). The reflectance spectra were normalized with respect to the reflectance from a gold film. Given that the transmittance was zero for the thick back reflector, the absorptivity spectra were calculated using the formula $A = 1 - R$, where A was the absorptivity, R was the reflectivity. The simulated and measured absorptivity spectra are shown in Figure 7e,f. The reflectance spectra were normalized with respect to the reflectance from a gold film. The four sensing elements exhibited absorptivity peaks of 0.92, 0.93, 0.85, and 0.87 at 3.32, 3.74, 4.06, and 4.51 µm, respectively, which were highly consistent with the corresponding simulated absorptivity peaks of 0.94, 0.99, 0.94, 0.98 at 3.33, 3.72, 4.05, and 4.50 µm, respectively. The precise pattern transfer in the fabrication process resulted in small shifts of only a few tens nm and slightly lower magnitudes of absorption peaks. The FWHMs of the measured absorptivity curves were 0.35, 0.45, 0.49, 0.68 µm, compared to those of the simulated curves, which were 0.30, 0.44, 0.47, 0.63 µm. Together with the defined patterns and uniformity observed in the SEM images, the excellent agreement between the simulated and experimentally measured absorptivity proved the quality of the optimized fabrication method.

The performance of the detector was evaluated by measuring its spectral responses to the IR radiation from a wavelength-tunable pulsed laser system in the range of 2.5–6.0 µm. The pulse width of the laser was 104 fs, and the repetition rate was 1 kHz. The IR laser beam was guided to the sensing area of the detector. The output electrical signal was amplified with a preamplifier and demodulated with a lock-in amplifier. The spectrally selective absorption of laser pulses at resonant wavelengths was due to the excitation of highly localized magnetic and electric dipole resonances, which was evidenced by the simulated field distributions (see Figure 2b–d). Such strong resonances efficiently confined the electromagnetic energy and provided sufficient time to convert it into resistive heat within the Al disks and the continuous Al back plate. Al_2O_3 underwent almost no loss in the mid-infrared region (see Figure 2f); that is, dielectric losses in the spacer layer were negligible in contrast to the plasmonic perfect absorbers operating in lower frequencies, such as in the microwave and terahertz region, in which the absorption was mainly due to dielectric losses. The heat at the continuous Al back plate that was associated with spectrally selective absorption was directly transferred to the zinc oxide thin film, which was among the materials that were electrically polarized due to their c-axis-oriented textured crystal structure [46–48]. The zinc oxide film consequently became electrically polarized with changes in the temperature, and the amount of electric charges was also proportional to the temperature fluctuation. The electrically polarized zinc oxide film hence produced a voltage between the Al and Pt electrodes. Pt was chosen as the bottom electrode because it has been reported that the hexagonal arrangement of the atoms in the (111) plane of Pt facilitated the nucleation and growth of hexagonal zinc oxide oriented along the [0001] direction [49]. In Figure 7g, the measured spectral responsivity curves of the four sensing elements are plotted. As seen from the results, the spectral responses of the four sensing elements were 125, 150, 126, 128 mV/W at 3.30, 3.71, 4.09, and 4.54 µm, respectively. The excellent measured spectral responses proved that the detector exhibited low thermal drift and thermal noise due to the low mass of the membrane-based design. The measured FWHMs of the spectral responses were 0.94, 1.02, 1.10, 1.20 µm, respectively, which were approximately two times broader than those of the absorptivity measured using the FTIR spectrometer. The broadened spectra could be partly attributed to the spectral bandwidth of the IR pulsed laser and the fluctuation of the incident angles and positions of the laser pulses. These experimental results clearly demonstrated that the proposed multispectral hybrid plasmonic–pyroelectric detectors can be fabricated and easily integrated into existing IR spectroscopic devices such as miniature NDIR sensors, chemical- and bio-sensors, photoacoustic imaging systems, and thermographic cameras. As an example, on-chip multi-spectral infrared detectors may offer an opportunity to develop miniaturized and robust photoacoustic computed tomography [50,51]. Specifically, in all-optical photoacoustic imaging systems, shifts in the wavelengths of light are able to be detected and analyzed using an array of multi-spectral micro-detectors.

4. Conclusions

We proposed and successfully fabricated and characterized a MEMS-based quad-wavelength hybrid plasmonic–pyroelectric infrared detector. Our device exploited the great tunability and high spectral resolution of patterned Al-disk-array/Al_2O_3/Al resonant perfect absorbers together with the MEMS-based pyroelectric platform. The absorption spectra of the plasmonic perfect absorbers were simulated and optimized using the FDTD and RCWA methods. The fabricated device demonstrated the spectrally selective detection of IR radiation at 3.3, 3.7, 4.1, and 4.5 µm, which are in the atmospheric window, with FWHMs of 0.94, 1.02, 1.10, 1.20 µm and responsivities of 125, 150, 126, 128 mV/W, respectively. These results demonstrate the feasibility and great potential of multispectral hybrid plasmonic–pyroelectric infrared detectors for a wide range of modern IR spectroscopic devices.

Author Contributions: Conceptualization T.Y., A.T.D., T.D.D., S.I., S.M., T.N.; methodology, T.Y., A.T.D., S.I., T.D.D., A.O., T.N.; software, A.T.D., T.D.D.; characterization, A.T.D., T.Y., T.D.D., Y.W.; writing original draft preparation, A.T.D., T.Y.; supervision, T.N.

Funding: This research was funded by Core Research for Evolutional Science and Technology (CREST) program (JPMJCR13C3) from the Japan Science and Technology Agency and the Kakenhi program 16H06364 and 16F16315, supported by Japan Society for the Promotion of Science.

Acknowledgments: The authors thank all the staff at Namiki Foundry and Nanofabrication Platform in National Institute for Materials Science for their kind supports in the device fabrication, especially T. Ohki and T. Sawada.

Conflicts of Interest: The authors declare no conflict of interest.

References

1. Ng, D.; Fralick, G. Use of a multiwavelength pyrometer in several elevated temperature aerospace applications. *Rev. Sci. Instrum.* **2001**, *72*, 1522–1530. [CrossRef]
2. Felice, R.A. The Spectropyrometer—A Practical Multi-wavelength Pyrometer. *AIP Conf. Proc.* **2003**, *684*, 711–716.
3. Boebel, F.G.; Möller, H.; Hertel, B.; Grothe, H.; Schraud, G.; Schröder, S.; Chow, P. In situ film thickness and temperature control of molecular beam epitaxy growth by pyrometric interferometry. *J. Cryst. Growth* **1995**, *150*, 54–61. [CrossRef]
4. Fu, T.; Liu, J.; Duan, M.; Zong, A. Temperature measurements using multicolor pyrometry in thermal radiation heating environments. *Rev. Sci. Instrum.* **2014**, *85*, 044901. [CrossRef]
5. Boboridis, K.; Obst, A.W. A High-Speed Four-Channel Infrared Pyrometer. *AIP Conf. Proc.* **2003**, *684*, 759–764.
6. Dinh, T.-V.; Choi, I.-Y.; Son, Y.-S.; Kim, J.-C. A review on non-dispersive infrared gas sensors: Improvement of sensor detection limit and interference correction. *Sens. Actuators B Chem.* **2016**, *231*, 529–538. [CrossRef]
7. Hodgkinson, J.; Smith, R.; Ho, W.O.; Saffell, J.R.; Tatam, R.P. Non-dispersive infra-red (NDIR) measurement of carbon dioxide at 4.2 μm in a compact and optically efficient sensor. *Sens. Actuators B Chem.* **2013**, *186*, 580–588. [CrossRef]
8. Hasan, D.; Lee, C.; Hasan, D.; Lee, C. Hybrid Metamaterial Absorber Platform for Sensing of CO_2 Gas at Mid-IR. *Adv. Sci.* **2018**, *5*, 1700581. [CrossRef]
9. Hodgkinson, J.; Tatam, R.P. Optical gas sensing: A review. *Meas. Sci. Technol.* **2013**, *24*, 012004. [CrossRef]
10. Tan, Q.; Zhang, W.; Xue, C.; Xiong, J.; Ma, Y.; Wen, F. Design of mini-multi-gas monitoring system based on IR absorption. *Opt. Laser Technol.* **2008**, *40*, 703–710. [CrossRef]
11. Wang, L.; Mizaikoff, B. Application of multivariate data-analysis techniques to biomedical diagnostics based on mid-infrared spectroscopy. *Anal. Bioanal. Chem.* **2008**, *391*, 1641–1654. [CrossRef]
12. Elliott, A.; Ambrose, E.J. Structure of Synthetic Polypeptides. *Nature* **1950**, *165*, 921–922. [CrossRef]
13. López-Lorente, Á.I.; Mizaikoff, B. Mid-infrared spectroscopy for protein analysis: Potential and challenges. *Anal. Bioanal. Chem.* **2016**, *408*, 2875–2889. [CrossRef]
14. Moore, D.J.; Sills, R.H.; Mendelsohn, R. Peroxidation of erythrocytes: FTIR spectroscopy studies of extracted lipids, isolated membranes, and intact cells. *Biospectroscopy* **1995**, *1*, 133–140. [CrossRef]
15. Wolkers, W.F.; Hoekstra, F.A. In situ FTIR Assessment of Desiccation-Tolerant Tissues. *Spectroscopy* **2003**, *17*, 297–313. [CrossRef]
16. Volkov, V.V.; Chelli, R.; Righini, R. Domain Formation in Lipid Bilayers Probed by Two-Dimensional Infrared Spectroscopy. *J. Phys. Chem. B* **2006**, *110*, 1499–1501. [CrossRef]
17. Marks, R.S.; Abdulhalim, I. *Nanomaterials for Water Management: Signal Amplification for Biosensing from Nanostructures*; Pan Stanford: New York, NY, USA, 2015; ISBN 9789814463478.
18. Nagao, T.; Han, G.; Hoang, C.; Wi, J.-S.; Pucci, A.; Weber, D.; Neubrech, F.; Silkin, V.M.; Enders, D.; Saito, O.; et al. Plasmons in nanoscale and atomic-scale systems. *Sci. Technol. Adv. Mater.* **2010**, *11*, 054506. [CrossRef]
19. Ariyawansa, G.; Rinzan, M.B.M.; Alevli, M.; Strassburg, M.; Dietz, N.; Perera, A.G.U.; Matsik, S.G.; Asghar, A.; Ferguson, I.T.; Luo, H.; et al. GaN/AlGaN ultraviolet/infrared dual-band detector. *Appl. Phys. Lett.* **2006**, *89*, 091113. [CrossRef]
20. Hudson, M.K.; Busch, K.W. Infrared emission from a flame as the basis for chromatographic detection of organic compounds. *Anal. Chem.* **1987**, *59*, 2603–2609. [CrossRef]
21. Liu, Z.; Kim, A.K. Review of Recent Developments in Fire Detection Technologies. *J. Fire Prot. Eng.* **2003**, *13*, 129–151. [CrossRef]

22. Romero, C.; Li, X.; Keyvan, S.; Rossow, R. Spectrometer-based combustion monitoring for flame stoichiometry and temperature control. *Appl. Therm. Eng.* **2005**, *25*, 659–676. [CrossRef]

23. Berni, J.; Zarco-Tejada, P.J.; Suarez, L.; Fereres, E. Thermal and Narrowband Multispectral Remote Sensing for Vegetation Monitoring From an Unmanned Aerial Vehicle. *IEEE Trans. Geosci. Remote Sens.* **2009**, *47*, 722–738. [CrossRef]

24. Meriaudeau, F. Real time multispectral high temperature measurement: Application to control in the industry. *Image Vis. Comput.* **2007**, *25*, 1124–1133. [CrossRef]

25. Gowen, A.A.; Tiwari, B.K.; Cullen, P.J.; McDonnell, K.; O'Donnell, C.P. Applications of thermal imaging in food quality and safety assessment. *Trends Food Sci. Technol.* **2010**, *21*, 190–200. [CrossRef]

26. Ünlü, M.S.; Strite, S. Resonant cavity enhanced photonic devices. *J. Appl. Phys.* **1995**, *78*, 607–639. [CrossRef]

27. Emsley, M.K.; Dosunmu, O.; Unlu, M.S. High-speed resonant-cavity-enhanced silicon photodetectors on reflecting silicon-on-insulator substrates. *IEEE Photonics Technol. Lett.* **2002**, *14*, 519–521. [CrossRef]

28. Boeberl, M.; Fromherz, T.; Schwarzl, T.; Springholz, G.; Heiss, W. IV-VI resonant-cavity enhanced photodetectors for the mid-infrared. *Semicond. Sci. Technol.* **2004**, *19*, L115–L117. [CrossRef]

29. Arnold, M.; Zimin, D.; Zogg, H. Resonant-cavity-enhanced photodetectors for the mid-infrared. *Appl. Phys. Lett.* **2005**, *87*, 141103. [CrossRef]

30. Wehner, J.G.A.; Musca, C.A.; Sewell, R.H.; Dell, J.M.; Faraone, L. Mercury cadmium telluride resonant-cavity-enhanced photoconductive infrared detectors. *Appl. Phys. Lett.* **2005**, *87*, 211104. [CrossRef]

31. Kishino, K.; Unlu, M.S.; Chyi, J.-I.; Reed, J.; Arsenault, L.; Morkoc, H. Resonant cavity-enhanced (RCE) photodetectors. *IEEE J. Quantum Electron.* **1991**, *27*, 2025–2034. [CrossRef]

32. Landy, N.I.; Sajuyigbe, S.; Mock, J.J.; Smith, D.R.; Padilla, W.J. Perfect Metamaterial Absorber. *Phys. Rev. Lett.* **2008**, *100*, 207402. [CrossRef] [PubMed]

33. Watts, C.M.; Liu, X.; Padilla, W.J. Metamaterial Electromagnetic Wave Absorbers. *Adv. Mater.* **2012**, *24*, OP98–OP120. [CrossRef] [PubMed]

34. Dao, T.D.; Ishii, S.; Yokoyama, T.; Sawada, T.; Sugavaneshwar, R.P.; Chen, K.; Wada, Y.; Nabatame, T.; Nagao, T. Hole Array Perfect Absorbers for Spectrally Selective Midwavelength Infrared Pyroelectric Detectors. *ACS Photonics* **2016**, *3*, 1271–1278. [CrossRef]

35. Liu, X.; Tyler, T.; Starr, T.; Starr, A.F.; Jokerst, N.M.; Padilla, W.J. Taming the Blackbody with Infrared Metamaterials as Selective Thermal Emitters. *Phys. Rev. Lett.* **2011**, *107*, 045901. [CrossRef] [PubMed]

36. Hao, J.; Wang, J.; Liu, X.; Padilla, W.J.; Zhou, L.; Qiu, M. High performance optical absorber based on a plasmonic metamaterial. *Appl. Phys. Lett.* **2010**, *96*, 251104. [CrossRef]

37. Korotchenkov, G.S. *Handbook of Gas Sensor Materials: Properties, Advantages and Shortcomings for Applications. New Trends and Technologies*; Springer: Berlin, Germany, 2013.

38. Esler, M.B.; Griffith, D.W.; Wilson, S.R.; Steele, L.P. Precision trace gas analysis by FT-IR spectroscopy. 1. Simultaneous analysis of CO_2, CH_4, N_2O, and CO in air. *Anal. Chem.* **2000**, *72*, 206–215. [CrossRef] [PubMed]

39. Dao, T.D.; Chen, K.; Ishii, S.; Ohi, A.; Nabatame, T.; Kitajima, M.; Nagao, T. Infrared Perfect Absorbers Fabricated by Colloidal Mask Etching of Al-Al$_2$O$_3$-Al Trilayers. *ACS Photonics* **2015**, *2*, 964–970. [CrossRef]

40. Synopsys's Rsoft. Available online: https://www.synopsys.com (accessed on 7 April 2019).

41. Malka, D.; Danan, Y.; Ramon, Y.; Zalevsky, Z.; Malka, D.; Danan, Y.; Ramon, Y.; Zalevsky, Z. A Photonic 1 × 4 Power Splitter Based on Multimode Interference in Silicon–Gallium-Nitride Slot Waveguide Structures. *Materials* **2016**, *9*, 516. [CrossRef]

42. Shoresh, T.; Katanov, N.; Malka, D. 1 × 4 MMI visible light wavelength demultiplexer based on a GaN slot-waveguide structure. *Photonics Nanostruct. Fundam. Appl.* **2018**, *30*, 45–49. [CrossRef]

43. Yokoyama, T.; Dao, T.D.; Chen, K.; Ishii, S.; Sugavaneshwar, R.P.; Kitajima, M.; Nagao, T. Spectrally Selective Mid-Infrared Thermal Emission from Molybdenum Plasmonic Metamaterial Operated up to 1000 °C. *Adv. Opt. Mater.* **2016**, *4*, 1987–1992. [CrossRef]

44. Khorasaninejad, M.; Zhu, A.Y.; Roques-Carmes, C.; Chen, W.T.; Oh, J.; Mishra, I.; Devlin, R.C.; Capasso, F. Polarization-Insensitive Metalenses at Visible Wavelengths. *Nano Lett.* **2016**, *16*, 7229–7234. [CrossRef] [PubMed]

45. Palik, E.D. *Handbook of Optical Constants of Solids*; Academic Press: Cambridge, MA, USA, 1998; ISBN 9780125444156.

46. Heiland, G.; Ibach, H. Pyroelectricity of zinc oxide. *Solid State Commun.* **1966**, *4*, 353–356. [CrossRef]

47. Hsiao, C.-C.; Huang, K.-Y.; Hu, Y.-C.; Hsiao, C.-C.; Huang, K.-Y.; Hu, Y.-C. Fabrication of a ZnO Pyroelectric Sensor. *Sensors* **2008**, *8*, 185–192. [CrossRef] [PubMed]

48. Hsiao, C.-C.; Yu, S.-Y.; Hsiao, C.-C.; Yu, S.-Y. Improved Response of ZnO Films for Pyroelectric Devices. *Sensors* **2012**, *12*, 17007–17022. [CrossRef] [PubMed]

49. Mirica, E.; Kowach, G.; Evans, P.; Du, H. Morphological Evolution of ZnO Thin Films Deposited by Reactive Sputtering. *Cryst. Growth Des.* **2003**, *4*, 147–156. [CrossRef]

50. Yao, J.; Xia, J.; Maslov, K.I.; Nasiriavanaki, M.; Tsytsarev, V.; Demchenko, A.V.; Wang, L.V. Noninvasive photoacoustic computed tomography of mouse brain metabolism in vivo. *Neuroimage* **2013**, *64*, 257–266. [CrossRef] [PubMed]

51. Wissmeyer, G.; Pleitez, M.A.; Rosenthal, A.; Ntziachristos, V. Looking at sound: Optoacoustics with all-optical ultrasound detection. *Light Sci. Appl.* **2018**, *7*, 53. [CrossRef] [PubMed]

Article

Efficient Fabrication Process of Ordered Metal Nanodot Arrays for Infrared Plasmonic Sensor

Masahiko Yoshino [1,*], Yusuke Kubota [1], Yuki Nakagawa [1] and Motoki Terano [2]

[1] Mechanical Engineering, School of Engineering, Tokyo Institute of Technology, Tokyo 152-8550, Japan; kubota.y.am@m.titech.ac.jp (Y.K.); nakagawa.y.aw@m.titech.ac.jp (Y.N.)

[2] Mechanical Systems Engineering, Faculty of Engineering, Okayama University of Science, Okayama 700-0005, Japan; m_terano@mech.ous.ac.jp

* Correspondence: myoshino@mes.titech.ac.jp; Tel.: +81-3-5734-2506

Received: 8 May 2019; Accepted: 4 June 2019; Published: 8 June 2019

Abstract: In this paper, a simple process to fabricate ordered Au nanodot arrays up to 520 nm in diameter that respond to infrared light is developed, and the feasibility of its application to infrared plasmonic sensors is shown. The developed process utilizes thermal dewetting to agglomerate a coated gold film into nanodots. It was difficult to produce large nanodots that responded to infrared light owing to dot separation. In this paper, therefore, the mechanism of dot agglomeration by thermal dewetting is studied via an experiment and theoretical model, and conditions to form single nanodots are clarified. Furthermore, Au nanodot arrays of 100 nm to 520 nm in diameter were fabricated by this process, and their absorption spectra were analyzed. In addition, an analysis of the change in the peak wavelength against the refractive index indicates the possibility of further improvement of the sensitivity of the infrared plasmon sensors.

Keywords: nanomaterials; plasmonics; nanodot array; thermal dewetting; absorbance spectrum; infrared plasmonic sensor

---

1. Introduction

Metal nano/micro structures smaller than the wavelength of light such as nanodot arrays generate localized surface plasmon resonance by incident light. These structures are expected to be applied to various photo devices, for example, plasmonic chemical sensors using metal nanodots, surface enhanced Raman scattering (SERS) substrates, and devices to enhance luminescence/absorption [1–11]. In particular, their applications to infrared devices have been attracting considerable attention because infrared light interacts with many substances, making it useful for material analysis. Several studies on infrared devices such as infrared band filters [12,13], infrared plasmonic sensors [14,15], heat radiation/absorption control [16,17], and infrared image sensing have been reported.

The electron beam lithography (EBL) process is considered to be suitable to fabricate nano/micro structures [18–22]; however, it requires expensive facilities and stringent control of process conditions, consequently increasing the product's cost. In addition, hazardous chemicals are necessary for the etching process, rendering this method dangerous in terms of environmental load. Therefore, an efficient and inexpensive method for manufacturing metallic nano/micro structures is required. To address the high product cost problem, many processes have been proposed, including nanoimprinting lithography [23,24], chemical stamp lithography, the porous anode alumina method [25,26], and colloidal lithography [6,27–30]. However, the capability of size control, precision of dimensions, and productivity are still insufficient for plasmonic sensor applications to be achieved.

The authors developed a simple and efficient manufacturing method for metal nanodot arrays by combining nano plastic-forming and thermal dewetting [31–35]. This method does not utilize EBL and it is not harmful to the environment. They fabricated nanodot arrays of hemisphere metal nanodots ranging from 75 nm to 200 nm in diameter. which exhibited an absorbance peak by localized surface plasmon resonance (LSPR) in the visible light range. It was shown that this peak wavelength could be controlled by adjusting the nano plastic-forming condition, making this method useful for efficient fabrication of plasmonic sensors. In addition, increments in sensitivity and SERS have been achieved by multilayering a metal nanodot array [36,37].

Owing to dot separation in thermal dewetting, these Au nanodot arrays were limited to a diameter of 200 nm or less, and exhibited plasmon resonance in the visible light range. Because of that, they could not be applicable to the infrared sensors that require larger arrays. Despite many studies on thermal dewetting instability, the separation mechanism of the coated metal film and the effects of process conditions on the dot size are not apparent [38–41]. To stably produce a nanodot array larger than 200 nm and determine optimal process conditions, it is necessary to clarify the principle of dot separation.

In this paper, the authors aim to clarify the principle of thermal dewetting dot separation in their developed process and show a possible application of Au nanodot arrays to infrared plasmon sensors. First, the experimental method and results are explained. Then, theoretical models are proposed to discuss characteristic phenomena of dot agglomeration by thermal dewetting. Through the discussion, the process conditions necessary to fabricate a nanodot array of target size are clarified. Finally, optical properties of the nanodot arrays are analyzed by a spectroscope. Results indicate that fabricated nanodot arrays have an absorption peak in the infrared range, and its peak wavelength depends on the refractive index of the surrounding liquid. There are several advantages of Au nanodot arrays when compared with other metal materials such as the lack of aging variation interference, because Au is hardly oxidized in the atmosphere. As a result, non-oxidation facilities such as vacuum chambers and inert gas chambers are not needed in its manufacturing, consequently reducing the product's cost. Although the proposed process is also applicable to other noble metals such as Ag and Pt, as previously reported, only Au is used as nanodot material in this paper. The methods and results reported can be also useful for fabrication of nanodot arrays of various metals.

2. Fabrication Process for Metal Nanodot Array

Figure 1 shows the fabrication method for a gold nanodot array. This process consists of three steps: (a) sputter coating of Au film, (b) groove grid patterning on the Au film, and (c) thermal dewetting to form a Au nanodot array [33]. In step (a), a quartz glass plate of 1 mm in thickness, finished by polishing to an optical flat, was used as a substrate. The quartz glass substrate was cleaned ultrasonically in an acetone bath and was then dried in air at ambient temperature. Next, the Au was coated on a quartz glass substrate by using a DC (direct current) sputter coater. The sputtering gas was Ar, and the pressure was 10 Pa. In step (b), a nanogroove grid was machined on Au thin film by repeating step feed and indentation motion of a knife edge too using the nano plastic-forming technique. Motion of the nano plastic-forming device was controlled by a computer, and a groove grid of arbitrary size could be produced. In step (c), the machined quartz glass substrate was annealed in atmosphere using an electric furnace. By this process, the Au thin films sectioned by the groove grid were aggregated into uniform nanodots that were aligned in an orderly fashion at uniform intervals.

Figure 2a shows the nano plastic-forming device. It consists of ultra-precision XY stages and a Z stage. Their feed resolutions are 1 nm for the X stage and 10 nm for the Y and Z stages. A specimen is mounted on the XY stage, and a knife edge tool is mounted on Z stage together with a load cell. The knife edge tool is indented onto the specimen by moving the Z stage downward. Figure 2b shows the knife edge tool made of a sharply polished (edge radius smaller than 30 nm) single crystal diamond. The edge's angle is 60° and the width of the diamond tool is 0.6 mm (more details of the equipment are explained in previous paper [32]). As shown in Figure 2c, a series of parallel nanogrooves are machined on the Au film by indenting the knife edge tool. For this experiment, the indentation load of the knife edge tool is set to 1.2 N. Then, the substrate is rotated 90° horizontally, and another series of parallel grooves are machined on the preformed grooves. Because all motion of the nano plastic forming device are controlled by a computer program, a nanogroove grid of an arbitrary size larger than 100 nm can be machined by this method. Table 1 lists the experimental conditions for examining the characteristics of the aggregation of Au nanodot arrays.

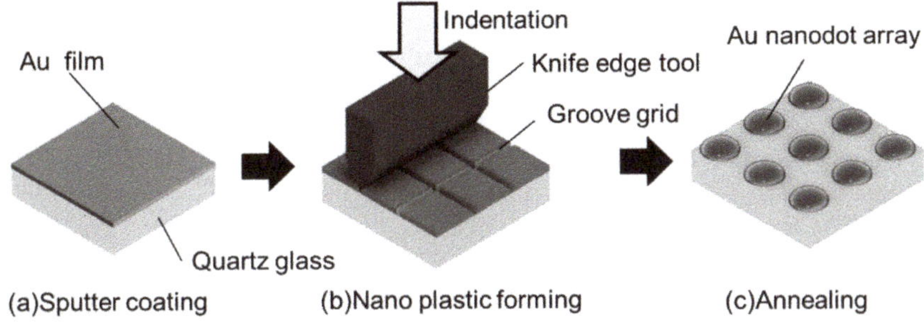

Figure 1. Schematic illustration of proposed fabrication process.

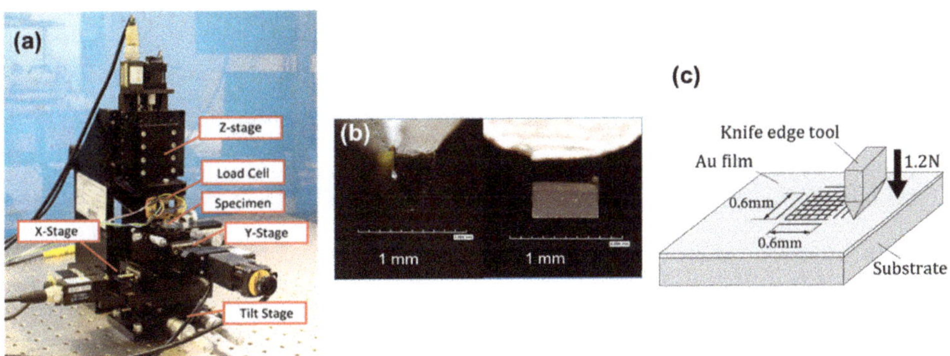

Figure 2. Nano plastic-forming device and knife edge tool made of single crystal diamond. (**a**) whole image of the device; (**b**) optical microscope image of the tool; (**c**) illustration for nanogroove fabrication method.

Table 1. Experimental conditions of first-layer nanodot array.

Thickness of Gold Film t, nm	Grid Size P, nm	Annealing Temperature T, °C	Annealing Time, min
5	50, 75, 100, 175, 250	700	10
10	250, 500, 1000	700	10
30	800, 900	1000	10
40	1000, 1200	1000	10

3. Experimental Results and Agglomeration Mechanism by Thermal Dewetting

3.1. Agglomeration of a Single Nanodot

Figure 3 shows scanning electron microscopy (SEM) images of the nanodot array. When the groove grid size and the thickness of the gold thin film are appropriate, the Au films in each grid are agglomerated into a single dot forming a uniform dot array with regular intervals (a,e). If the size and thickness are not appropriate, the gold thin film in the groove grid is separated and agglomerated into multiple dots (c,d). In addition, the size of the nanodots changes based on the groove grid size. These dots are well aligned in a lattice pattern, and most of them are formed into an almost hemispherical shape, while some of them are distorted into an ellipsoidal shape. Although the detail of this distortion is not yet fully justifiable, insufficient agglomeration is one of its main causes. It can be reduced by increasing annealing temperature or annealing time. Figure 3d illustrates an example where dot agglomeration is not complete, but the gold film is separated into many pieces, which will be agglomerated into multiple dots.

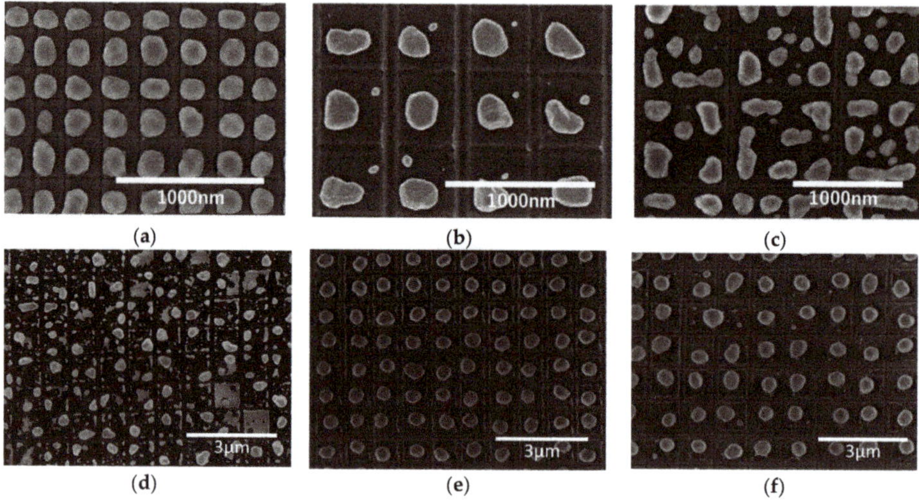

Figure 3. Field-emission scanning electron microscope (FE-SEM) micrographs of gold nanodot array generated by annealing; (**a**) t = 10 nm, P = 250 nm, T = 700 °C; (**b**) t = 10 nm, P = 500 nm, T = 700 °C; (**c**) t = 10 nm, P = 1000 nm, T = 700 °C; (**d**) t = 17 nm, P = 1000 nm, T = 700 °C; (**e**) t = 40 nm, P = 1000 nm, T = 1000 °C; (**f**) t = 40 nm, P = 1200 nm, T = 1000 °C.

Figure 4 shows the variation of the average nanodot diameter with respect to the size of the groove grid. In this figure, the solid symbols indicate the single-dot mode and the open symbols indicate the multiple-dot mode. The vertical bars represent the standard deviation of the dot diameter. Five to 10 specimens were tested on the same condition. In the single-dot mode, the average diameter of the nanodots increases with an increase in the grid size, as shown by the solid symbols. However, in the multiple-dot mode, the average diameter of the divided dots is small, and the standard deviation is large.

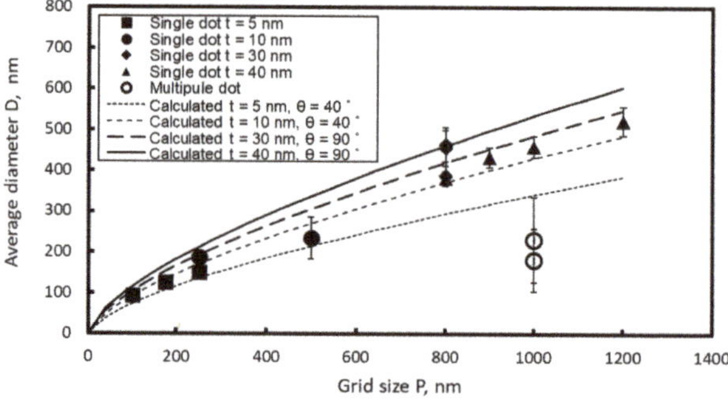

Figure 4. Variation of average nanodot size and its standard deviation against grid size.

In the single-nanodot mode, the diameter of the nanodot is calculated based on the volume preservation in the thermal dewetting process. The volume of a hemisphere nanodot is equivalent to the volume of the square film of a groove grid, as shown in the model of Figure 5.

$$\frac{\pi}{24}\frac{D^3}{\sin^3\theta_c}\left(2 - 3\cos\theta_c + \cos^3\theta_c\right) = P^2 t \tag{1}$$

The term on the left is the volume of a dot whose contact angle is θ_c. When the agglomeration is completed, θ_c is determined by the balance of the surface energy of the substrate γ_S, the surface energy of the nanodot γ_M, and the interface energy between the metal film and the substrate γ_I. This is expressed by Young's equation [42]:

$$\gamma_I - \gamma_s = -\gamma_M \cos\theta_c \tag{2}$$

From Equation (1), the diameter of the nanodot is obtained as follows:

$$D = \left\{\frac{24P^2 t}{\pi(2 - 3\cos\theta_c + \cos^3\theta_c)}\right\}^{\frac{1}{3}} \sin\theta \qquad (0° < \theta_c \leq 90°) \tag{3}$$

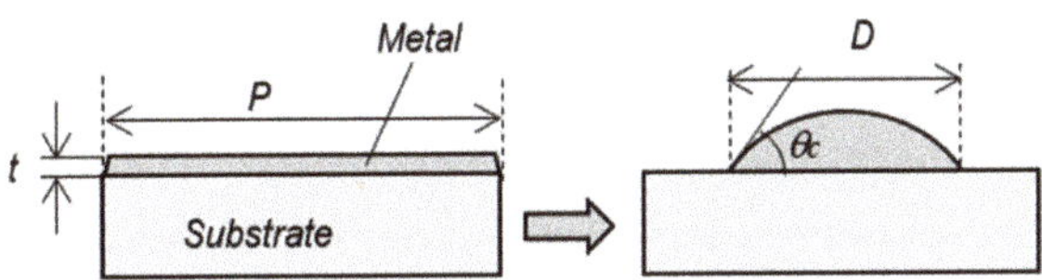

Figure 5. Geometrical change of a nanodot by thermal dewetting.

The curves in Figure 4 represent variations in D calculated by Equation (1). In this equation the dot diameter depends on the grid size and the contact angle; dot diameter becomes larger when the contact angle becomes smaller. In Figure 4, the contact angles were adjusted so that the calculated curves agreed with the experimental data: the contact angle $\theta c = 40°$ when $t = 5$ nm and $t = 10$ nm ($T = 700$ °C), and the contact angle $\theta c = 90°$ at $t = 30$ nm ($T = 1000$ °C). It is confirmed that the model explains variation of the nanodot diameter of single mode well. Although this calculation model assumes that all nanodots are agglomerated into a hemispherical form with equal contact angles, the actual nanodots are distorted as seen in Figure 2, and the contact angles varies along the circumference of a dot. Thus, the calculated contact angles do not represent the correct contact angles. However, the contact angles of 40° and 90° determined from the curve fitting can be considered representing the average contact angle of actual nanodots, including some 10° of error.

Difference between the contact angles of both conditions is attributed to the difference in the annealing temperature. Because thermal dewetting is attributed to atomic diffusion, and the diffusion coefficient depends on temperature, the agglomeration rate increases with annealing temperature. It is also known that the annealing temperature affects the surface energy of the gold and the quartz glass, and also the interface energy between the Au nanodots and quartz glass substrate. Because the contact angle of the Au nanodot depends on balance of these energies, the annealing temperature affects the aggregation process and final geometry of the Au nanodots. However, the contact angle is also affected by atmosphere gas, and influence of annealing temperature is not clear. Thus, we do not discuss the effect of the annealing temperature on the contact angle of the nanodots, but regard the contact angle obtained from the curve fitting in Figure 4 as the practical value of the average contact angle.

The agglomeration is driven by a reduction in the total free energy of the substrate/metal system. The total free energy of the substrate/metal system before annealing G_1 is calculated as follows:

$$G_1 = P^2\gamma_M + P^2\gamma_I = P^2(\gamma_M + \gamma_I) \tag{4}$$

The total free energy of a system in the agglomeration process is expressed as the summation of the surface energy of a dot, the interfacial energy, and the surface energy of the exposed substrate.

$$G_2 = S_M\gamma_M + S_I\gamma_I + \left(P^2 - S_I\right)\gamma_S \tag{5}$$

where S_M is the area of the dot surface, and S_I is the area of the interface between the dot and the substrate. They are calculated by the following equations:

$$S_M = \frac{\pi D^2}{2}\frac{1 - \cos\theta_c}{\sin^2\theta_c} \tag{6}$$

$$S_M = \frac{\pi D^2}{4} \tag{7}$$

Using Equations (2), (6), and (7), the total free energy is obtained as follows:

$$G_2{}^{single} = \frac{\pi D^2}{4}\frac{2 - \cos\theta_c - \cos^2\theta_c}{1 + \cos\theta_c}\gamma_M + P^2\gamma_S \tag{8}$$

Figure 6 shows the calculated variation of the free energies G_1 and $G_2{}^{single}$. As for the calculation, the grid size P was 1000 nm, and the contact angle θ_c was 90°. The surface energy of Au and SiO_2 referred to in the literature was adopted as γ_M and γ_I, respectively. It was shown that the free energy after dewetting $G_2{}^{single}$ was smaller than that before dewetting G_1 and nanodots are formed when the thickness of the metal film was small enough. In addition, this equation indicates possibility that agglomeration does not occur when the metal film is too thick.

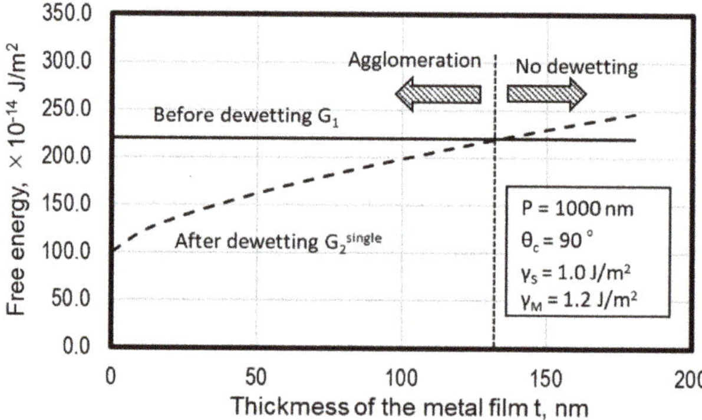

Figure 6. Variation of free energies G_1 and $G_2{}^{single}$ as calculated by Equation (4) and (8).

3.2. Agglomeration Mechanism of Multiple Nanodots

In case of very thin Au film, the film separated and agglomerated into multiple nanodots. This is considered to be the same phenomenon as the normal thermal dewetting process reported by Cheynis et al. [43] because the sectioned square region was relatively broad compared to the film thickness. Danielson [41] explained the agglomerated dot size based on the Rayleigh instability mechanism. However, the Rayleigh instability is a dynamic vibration phenomenon of a stream of falling water [44] and is not suitable for thermal dewetting, which is attributed to the diffusion mechanism.

In a typical thermal dewetting process, many microdefects in the metal film [45] are widened by solid diffusion and become voids. These voids grow and connect to adjacent voids owing to thermal dewetting, and create isolated metal islands. Finally, the isolated metal islands agglomerate into hemispherical dots. However, in the experiments in Figure 2, the area of agglomeration is restricted by the groove grids. For dot separation to occur, it is necessary that the dots do not contact each other after aggregation. Here, a case is assumed in which two dots of diameters D_1 and D_2 are formed in a square grid, as shown in Figure 7.

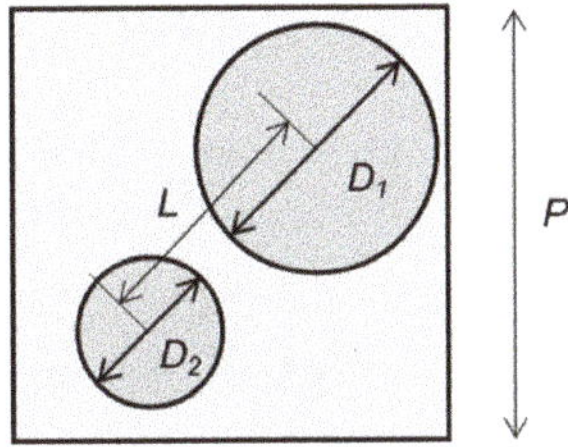

Figure 7. Two nanodots formed in a square grid.

Because the sum of both volumes is equal to the volume of the coated metal film of thickness t,

$$V_1 + V_2 = P^2 t \tag{9}$$

where V_1 and V_2 are the volumes of these dots. By expressing the ratio of D_1 and D_2 by k, i.e., $D_2 = kD_1$. the diameters of these dots are calculated by the following equation:

$$D_2 = kD_1 = k\left\{ \frac{24 sin^3\theta}{\pi} \frac{P^2 t}{(2 - 3\cos\theta + cos^3\theta)(1 + k^3)} \right\}^{\frac{1}{3}} \quad 0° < \theta \le 90° \tag{10}$$

If the distances of the dot centers are shorter than the sum of the radiuses of both dots, then the two dots come in contact and aggregate into one dot owing to the surface tension of the metal. Therefore,

$$\frac{(D_1 + D_2)}{2} = \frac{(D_1 + kD_1)}{2} = \frac{D_1}{2}(1 + k) > L \tag{11}$$

is the condition of a metal film in a square grid to be agglomerated in a single dot.

If the center of a dot is too close to the boundary, the outer periphery of the dot overlaps the groove, and it cannot be formed into a round dot. Thus, the center of the dot must be apart from the boundary by more than the distance of the radius. Therefore, the region in which a dot can exist is a square area of $L = P - \frac{D_1}{2} - \frac{D_2}{2}$ on each side. Because the average distance between two points randomly dispersed in a square area is expressed as $\mu L \cong 0.521L$ (see Appendix A), Equation (11) yields

$$\frac{D_1}{2}(1 + k) > \mu\left(P - \frac{D_1}{2} - \frac{D_2}{2}\right) \tag{12}$$

Substituting Equation (10) into Equation (12), we obtain the following inequation:

$$\frac{t}{P} > (2 - 3\cos\theta + cos^3\theta) \frac{\pi}{3 sin^3\theta} \frac{\mu^3}{(1 + \mu)^3} \frac{(1 + k^3)}{(1 + k)^3} \tag{13}$$

Hereinafter, t/P is called the critical film thickness ratio. It is considered that a single dot is generated when t/P is larger than the critical film thickness ratio, and multiple dots are generated when t/P is smaller than the critical film thickness ratio.

Figure 8 shows the variation of critical film thickness ratio calculated by Equation (13) (curve) and the experimental data in which single (solid symbols) and multiple (open symbols) dots were formed, against the contact angle. For reference, a contact error of range ±10° is also displayed. The solid line lies near the boundary between the white circles and the black circles; therefore, the model correctly explains the condition of the single-dot mode and multiple-dot mode the error associated with the prediction is considered due to the distortion of nanodots. Even when there are more than three dots, the critical film thickness is determined by a similar condition that dot diameter must be larger than the average of minimum distance between dots. Although details of calculations are not shown here, the obtained result proves that the critical film thickness ratio decreases as the number of dots in the lattice increases.

Figure 8 also indicates that the critical film thickness ratio increases with the contact angle. A single nanodot correctly aligned as the groove grid can easily be obtained in case of a substrate and metal with small contact angle. Although it is difficult to control the contact angle because it depends on the combination of materials and the annealing conditions, the dot size can be controlled by adjusting the film thickness ratio to a range larger than critical film thickness ratio.

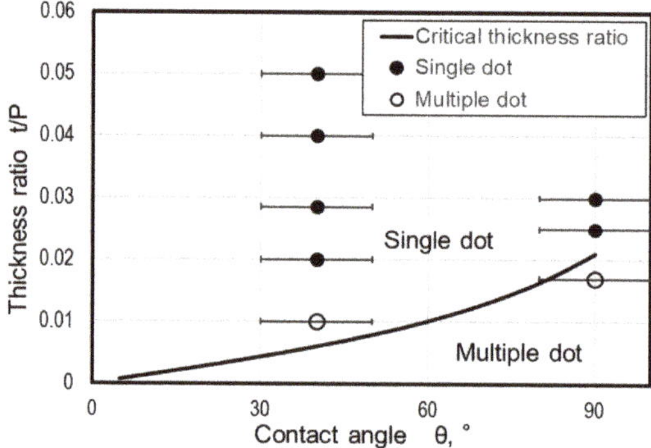

Figure 8. Comparison between theoretical value of critical film thickness ratio and experimental value of dot formation.

The total free energy of the dot/substrate system of the multiple-dot mode is also calculated by Equation (5), where S_M and S_I are calculated by the following equations:

$$S_M = \frac{\pi\left(D_1{}^2 + D_2{}^2 + \cdots + D_N{}^2\right)}{2}\frac{1 - \cos\theta_c}{\sin^2\theta_c} = \frac{\pi\sum_{i=1}^{N} D_i{}^2}{2}\frac{1 - \cos\theta_c}{\sin^2\theta_c} \tag{14}$$

$$S_I = \frac{\pi\left(D_1{}^2 + D_2{}^2 + \cdots + D_N{}^2\right)}{4} = \frac{\pi\sum_{i=1}^{N} D_i{}^2}{4} \tag{15}$$

where N is the number of dots in a square grid area. The total free energy of the system is obtained as follows:

$$G_2{}^{multiple} = \frac{\pi\sum_{i=1}^{N} D_i{}^2}{4}\frac{2 - \cos\theta_c - \cos^2\theta_c}{1 + \cos\theta_c}\gamma_M + P^2\gamma_S \tag{16}$$

Figure 9 shows the calculated variations of the free energy against the number of dots, where all of the dot sizes are assumed to be equal and the material data of Au and quartz glass are adopted. It is reasonable that the free energy increases with the number of nanodots. It is also found that the free energy increases with an increase in the metal film thickness. This is because an increase in film thickness results in an increase in the dot diameter and thus an increase in S_M of Equation (11). When the film becomes thicker, the increase in free energy against the increase in the dot number becomes larger and, therefore, a thick film is easier to agglomerate to a large dot.

Because a lower free energy increases the possibility of the phenomenon's occurrence, nanodots are agglomerated into a single dot only if a reduction in the free energy governs the dot agglomeration. However, the aggregation of the dots results from the solid diffusion of atoms in the metal film, and it is difficult for an aggregation pattern that needs a long migration of atoms to occur. In addition, it is known that a large number of microdefects in the metal film grow into many holes, and they separate the metal film by thermal dewetting. Therefore, it is understood that the coated metal film is agglomerated into multiple dots if there is enough space between dots, but in the case of the sectioned metal film of our process, the number of dots is restricted by the condition of the critical film thickness ratio in Equation (13).

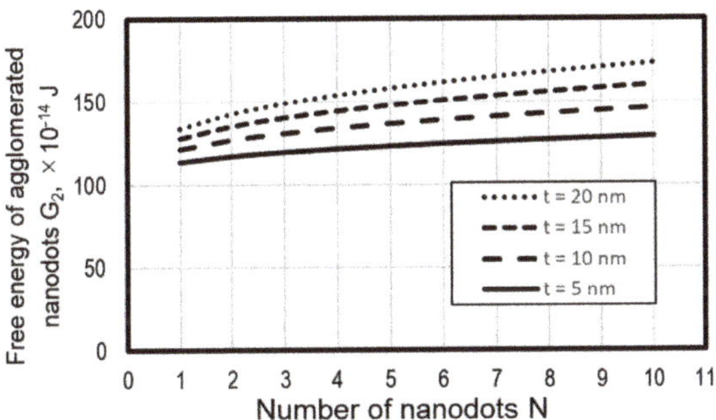

Figure 9. Variation of free energy of separated dots against the number of dots in a square grid.

4. Optical Properties of Au Nanodot Arrays

The absorbance spectra of the fabricated Au nanodot arrays were analyzed by using a fiber-type infrared (IR) spectrometer whose area of analysis was almost 0.5 mm in diameter. In addition, three kinds of liquid, i.e., water, acetone, and kerosene, were dropped on nanodot arrays, and the absorbance spectra were analyzed to evaluate the influence of the refractive index. Figure 10 shows the absorbance spectrum of an Au nanodot array measured by a fiber-type spectrophotometer. In the figure, a clear peak is observed in the near-infrared region, which is considered to be owing to localized plasmon resonance. It is found that the peak wavelength increases with increase of the dot diameter.

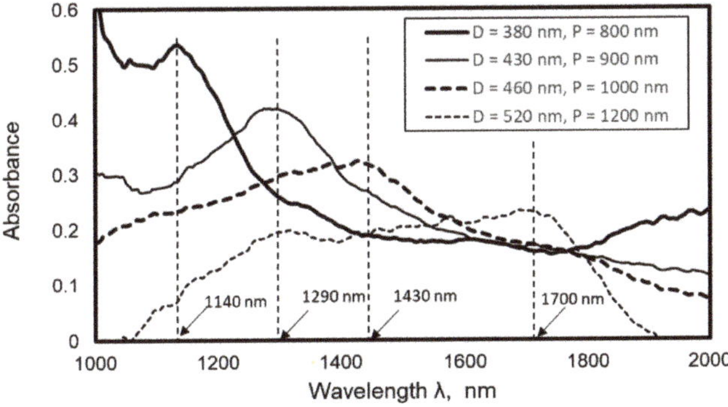

Figure 10. Influence of average dot diameter on absorbance spectrum of nanodot arrays.

Figure 11 shows the change in the peak wavelength with respect to the average diameter. Solid circles represent data of Figure 10 and solid triangles represent data obtained from a previous report [35]. It is found that the peak wavelength increases almost linearly with an increase in the average dot diameter D.

The solid line in Figure 11 is the modeled variation of peak wavelength of the absorbance spectrum based on the Kuwata's method [46]. In this model, the extinction cross section of an ellipsoidal nanodot, whose radius is denoted by a, b, and c, is calculated by

$$\sigma_{ext} = k \cdot Im(\alpha) \tag{17}$$

where k is the wave number of the incident light calculated by

$$k = \frac{2\pi (\varepsilon_m)^{\frac{1}{2}}}{\lambda} \tag{18}$$

ε_m is the dielectric constant around the nanodot, which is assumed as the average of the relative permittivity of quartz glass substrate and air [47]. λ is the wavelength of the incident light. α is the polarizability expressed by the following equation [46]:

$$\alpha = \frac{V}{\left(L + \frac{\varepsilon_m}{\varepsilon - \varepsilon_m}\right) + A(L)\varepsilon_m x^2 - i\frac{4\pi^2 \varepsilon_m^{3/2} V}{\lambda^3} + B(L)\varepsilon_m^2 x^4} \tag{19}$$

where ε is the complex dielectric constant of the nanodot metal, i.e., Au [48], x is the size parameter of an ellipsoidal nanodot ($x = \pi a / \lambda$), and V is the volume of the ellipsoidal nanodot. In the case of an oblate spheroid (a = b = 2c, where c is the radius to incident light direction), the geometrical factor L is calculated by the following equation [49]:

$$L = \frac{1}{2e^2}\left(\frac{1 - e^2}{e^2}\right)^{\frac{1}{2}}\left\{\frac{\pi}{2} - \tan^{-1}\left(e\left(\frac{1 - e^2}{e^2}\right)^{\frac{1}{2}}\right)\right\} - \frac{1}{2}\left(\frac{1 - e^2}{e^2}\right) \tag{20}$$

where e is the eccentricity.

$$e^2 = 1 - \frac{c^2}{a^2} \tag{21}$$

A(L) and B(L) in equation (16) are calculated by the following equations

$$A(L) = -0.4865L - 1.046L^2 + 0.8148L^3 \tag{22}$$

$$B(L) = 0.01909L + 0.1999L^2 + 0.6077L^3 \tag{23}$$

Using these equations, absorbance spectra of an oblate spheroid of various sizes is simulated and compared with the measured peak wavelength of the absorbance spectra. The simulation data in Figure 11 agrees well with the experimental data; thus, the absorbance peak found in Figure 10 can be attributed to the localized surface plasmon resonance. The residual difference between the simulation data and experimental data is due to the difference of dot shape because the actual Au dot fabricated by the experiment was not an oblate spheroid but a distorted hemisphere.

Another important conclusion from the theoretical model is that the absorbance peak wavelength due to the localized surface plasmon resonance also depends on the shape of the dot, i.e., the ratio of a, b, and c. As a result, the width of the absorbance spectrum peak spreads and peak wavelenght becomes unclear. Thus, it is important to reduce distortion of shape and size dispersion of nanodots to improve sharpness of the absorbance spectrum peak and accuracy of peak wavelength. If a sharp absorbance peak can be realized by this process, it can be used as an infrared filter. In order to detect a signal of a target wavelength with high sensitivity in material analysis by infrared (e.g., Fourier transform infrared (FT-IR)), it is necessary to reduce signal noise. An infrared filter of the Au nanodot array with required absorbance peak wavelength can be made by controlling the Au dot size by adjusting the nano groove grid size and the Au film thickness. Using this Au nanodot filter, noise can be appropriately reduced and the target signal can be successfully measured. Furthermore, the Au nanodot arrays can be utilized as an absorber of heat radiation energy.

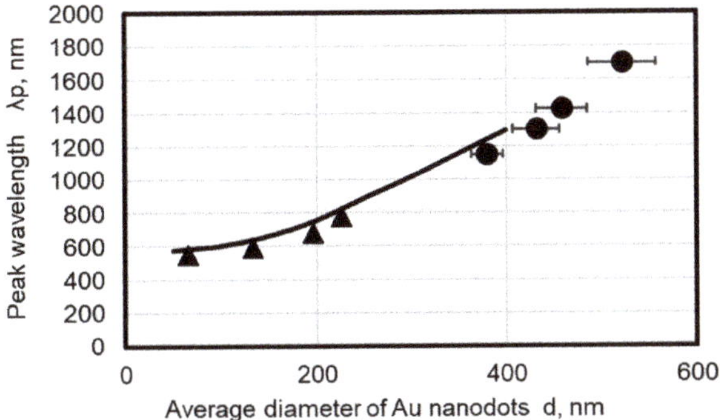

Figure 11. Change in peak wavelength against average dot diameter. ●: data obtained from Figure 10, ▲: data referred from a previous report [32], curve: calculated date.

Figure 12 shows the effects of the refractive index on the normalized absorbance spectra of a nanodot array of 459 nm in diameter, where water (refractive index n = 1.33), acetone (n = 1.36), and kerosene (n = 1.45) were dropped; data without liquid specimen, i.e., air (n = 1) are also shown. It is apparent that the peak of the absorbance spectrum shifted to the red side when the refractive index was modified by the dropped liquid. This characteristic is utilized in plasmonic sensors; the refractive index of a liquid specimen can be evaluated by a shift in the peak wavelength, which can be measured by the simple setup of a fiber-type spectroscope. This technology can be also applied to sensing of biomolecules such as protein or viruses.

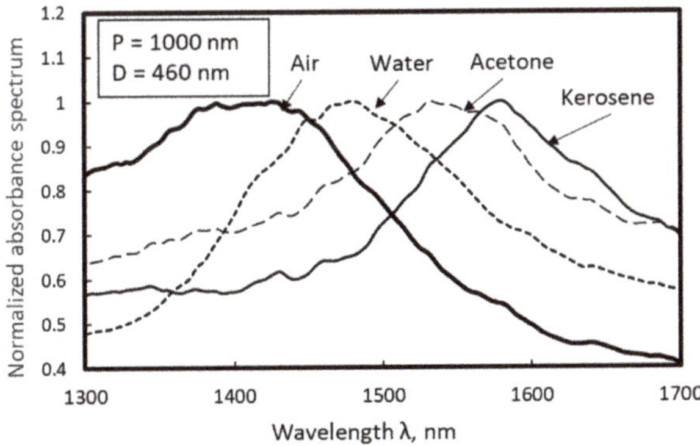

Figure 12. Effects of refractive index on absorbance spectra of nanodot array of 460 nm in diameter.

Figure 13 shows the variation in the peak wavelength based on the refractive index of the dropped liquid. The gradient of the approximated line (m-value) is also denoted in the graph. The m-value indicates the change in the peak wavelength of the absorbance spectrum (nm) against the change in the refractive index of the dropped liquid (RIU, refractive index unit). From Figure 13, it is found that the peak wavelength linearly increases with the dropped liquid's refractive index. It is also found that the m-value depends on the average dot diameter D.

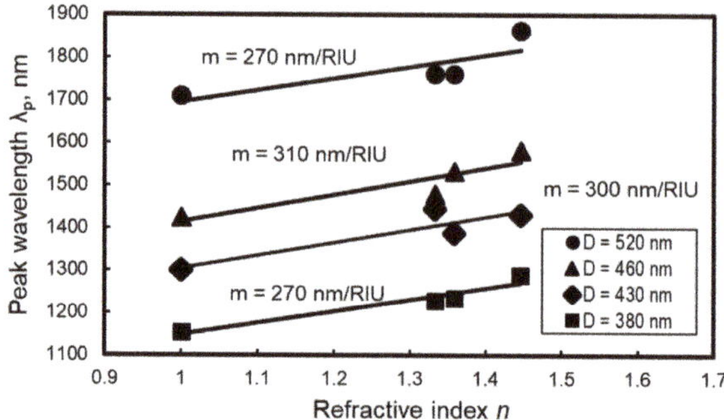

Figure 13. Change in peak wavelength against refractive index.

Figure 14 plots the m-value against the average dot diameter. The solid circles are data of m-values obtained from Figure 13, and the solid triangles are data of m-values in the visible light range referred from a previous report [34]. The figure shows that m-values obtained from Figure 13 are higher than those of the previous report, indicating that larger dots exhibit a higher m-value and a higher m-value can be obtained in the infrared range than in the visible light range. The line in Figure 14 represents the peak wavelength's variation in simulated absorbance spectra using Kuwata's method. The simulation results present a peak wavelength increase with dot diameter. This behavior is similar to the experimental data; however, the calculated m-value is almost twice as large. This is attributed to the difference in shape and the difference in the layout of the dot between the actual nanodots and the calculated model because actual nanodots are distorted hemispheres on the quartz glass. Nevertheless, the fundamental LSPR properties can still be explained by Kuwata's method. It is apparent that larger nanodots generate a higher m-value that is more likely to be obtained in the infrared range than in the visible light range. As the proposed fabrication process is able to control the size and geometry of the nanodot array, the sensitivity of plasmon sensors can be improved by developing suitable dot geometry for analysis by infrared light.

The nanodot array is expected to be applied in a chemical sensor that can analyze the component of a liquid specimen by measuring the refractive index based on the change in the absorption spectrum peak wavelength. To develop a sensitive chemical sensor, a higher m-value is preferable; therefore, a large nanodot array with a spectral peak in the infrared region is advantageous. Furthermore, a sharp absorption spectrum peak is preferable, and it is required to fabricate uniform nanodot arrays with small dispersion in dot diameter and little distortion in shape. For this purpose, further understanding of the thermal dewetting mechanism and development of control technology of the agglomeration process is necessary.

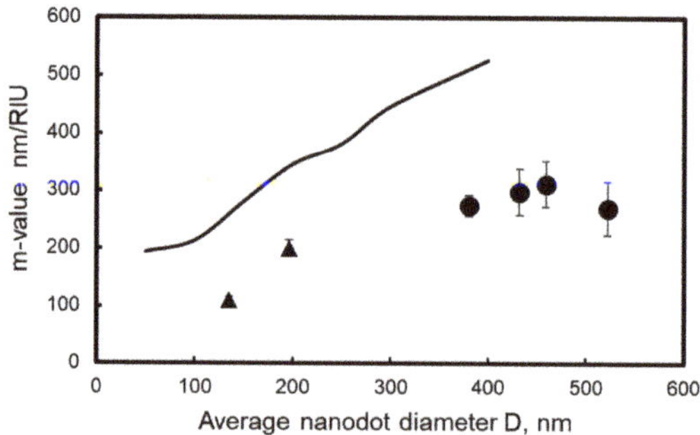

Figure 14. Change in sensitivity against average dot diameter. ●: data obtained from Figure 13, ▲: data referred from a previous report [35], curve: calculated date.

5. Conclusion

This study examined a simple and low-cost process for the fabrication of uniform metal nanodot arrays by combining nano plastic forming and thermal dewetting by annealing. The process conditions were clarified to produce relatively large nanodot arrays with a diameter up to 520 nm; these arrays could be useful for infrared light analysis.

The original process proposed faced a problem because the metal film had been separated into multiple dots during the thermal dewetting as the size of the nanogroove grid became large. In this study, therefore, a geometrical model to explain the mechanism of dot aggregation and separation by thermal dewetting was developed, and conditions for aggregation into single dots on each grid were clarified. The model was validated by comparing the results with the experimental data, enabling the fabrication of Au nanodot arrays of various dot sizes (from 100 nm to 520 nm). Although problems still remain with uniformity, distortion, and dispersion in size and geometry of nanodots, the feasibility of the process was confirmed.

Furthermore, it was shown that an Au nanodot array produced by this process exhibits an absorbance spectrum in the visible light to infrared range peak owing to the localized plasmon resonance. Because the peak wavelength of the absorbance spectrum depends on the average dot diameter, the proposed technology is effective to develop optical devices with various absorbance peaks. In addition, it is important to reduce the distortion of dots to obtain a unique absorbance peak as the distortion of the dot influences the peak wavelength.

Moreover, it was shown that the peak wavelength also depends on the liquid's refractive index dropped on the nanodots. A shift in the peak wavelength against change in the refractive index, i.e., m-value, was found to increase with average dot diameter. Because a high m-value is preferable for sensor sensitivity improvement, a large nanodot array with a spectral peak in the infrared region is advantageous for developing a highly sensitive sensor. The proposed fabrication process is, therefore, expected to improve the sensitivity of plasmon sensors by improving the dot geometry suitable for analysis by infrared light.

Author Contributions: Conceptualization, M.Y.; methodology, M.Y., M.T. and Y.N.; investigation, Y.K., Y.N. and M.T.; writing, M.Y., Y.K., Y.N. and M.T.; supervision, M.Y.; project administration, M.Y.; funding acquisition, M.Y.

Funding: This research was supported by a Grant-In-Aid for Scientific Research (A) (grant number 17H01240) of Japan Society for the Promotion of Science (JSPS).

Appendix A

The average distance between two points randomly dispersed in a unit square is calculated by the following equation:

$$\mu = \int_0^1 \int_0^1 \int_0^1 \int_0^1 \left\{ \sqrt{(x_1 - x_2)^2 + (y_1 - y_2)^2} \right\} dx_1 dx_2 dy_1 dy_2$$
$$= \frac{1}{15}\left\{2 + \sqrt{2} + 5\mathrm{arcsinh}(1)\right\} \cong 0.521$$

In the case that more than three points exist in a unit square, the average of the minimum points' distance is obtained by a numerical simulation in which a determined number of points were randomly distributed in a unit square by a random function, and then the minimum distance between them was calculated. By repeating this calculation more than 1,000,000 times, the average minimum distance was obtained. Table A1 shows the average of the minimum distances and standard deviations against the number of points used.

Table A1. Average of the minimum distances and standard deviations against the number of points.

Number of Points	Average of Minimum Distance	Standard Deviation
2	0.52106	0.24783
3	0.30556	0.16004
4	0.21179	0.10976
5	0.16233	0.08362
6	0.13189	0.06816
7	0.11103	0.05737
8	0.09583	0.04959
9	0.08436	0.04367
10	0.07538	0.03904

References

1. Hossain, M.K.; Huang, G.G.; Kaneko, T. Characteristics of surface-enhanced Raman scattering and surface-enhanced fluorescence using a single and a double layer gold nanostructure. *Phys. Chem. Chem. Phys.* **2001**, *11*, 7484–7490. [CrossRef] [PubMed]
2. Lahav, M.; Vaskevich, A.; Rubinstein, I. Biological sensing using transmission surface plasmon resonance spectroscopy. *Langmuir* **2004**, *20*, 7365–7367. [CrossRef] [PubMed]
3. Leebeeck, A.D.; Kumar, L.K.S.; Lange, V.; Sinton, D.; Gordon, R.; Brolo, A.G. On-chip surface-based detection with nanohole arrays. *Anal. Chem.* **2007**, *79*, 4094–4100. [CrossRef] [PubMed]
4. Gupta, G.; Tanaka, D.; Ito, Y.; Shibata, D.; Shimojo, M.; Furuya, K.; Mitsui, K.; Kajikawa, K. Absorption spectroscopy of gold nanoisland films: Optical and structural characterization. *Nanotechnology* **2009**, *20*, 025703. [CrossRef] [PubMed]
5. Mader, M.; Perlt, S.; Höche, T.; Hilmer, H.; Grundmann, M.; Rauschenbach, B. Gold nanostructure matrices by diffraction mask-projection laser ablation: Extension to previously inaccessible substrates. *Nanotechnology* **2010**, *21*, 175304. [CrossRef]
6. Yang, S.; Xu, F.; Ostendorp, S.; Wilde, C.; Zhao, H.; Lei, Y. Template-confined dewetting process to surface nanopatterns: Fabrication, structural tunability, and structure-related properties. *Adv. Funct. Mater.* **2011**, *21*, 2446–2455. [CrossRef]
7. Smitha, S.L.; Gopchandran, K.G.; Ravindran, T.R.; Prasad, V.S. Gold nanorods with finely tunable longitudinal surface plasmon resonance as SERS substrates. *Nanotechnology* **2011**, *22*, 265705. [CrossRef]
8. Hong, Y.; Huh, Y.M.; Yoon, D.S.; Yang, J. Nanobiosensors based on localized surface plasmon resonance for biomarker detection. *J. Nanomater.* **2013**, *2012*, 759830. [CrossRef]
9. Jiang, R.; Chen, H.; Shao, L.; Li, Q.; Wang, J. Unraveling the evolution and nature of the plasmons in (au core)–(ag shell) nanorods. *Adv. Opt. Mater.* **2012**, *24*, 200–207. [CrossRef]
10. Li, M.; Cushing, S.K.; Zhang, J.; Lankford, J.; Aguilar, Z.P.; Ma, D.; Wu, N. Shape-dependent surface-enhanced Raman scattering in gold–Raman-probe–silica sandwiched nanoparticles for biocompatible applications. *Nanotechnology* **2012**, *23*, 115501. [CrossRef]
11. Sun, H.; Yu, M.; Wang, G.; Sun, X.; Lian, J. Temperature-Dependent Morphology Evolution and Surface Plasmon Absorption of Ultrathin Gold Island Films. *J. Phys. Chem. C* **2012**, *116*, 9000–9008. [CrossRef]
12. Pinhas, H.; Malka, D.; Danan, Y.; Sinvani, M.; Zalevsky, Z. Design of fiber-integrated tunable thermo-optic C-band filter based on coated silicon slab. *J. Eur. Opt. Soc.-Rapid Publ.* **2017**, *13*. [CrossRef]
13. Zhang, B.; Zhao, Y.; Hao, Q.; Kiraly, B.; Khoo, I.-C.; Chen, S.; Huang, T.J. Polarization-independent dual-band infrared perfect absorber based on a metal-dielectric-metal elliptical nanodisk array. *Opt. Express* **2011**, *19*, 15221–15228. [CrossRef] [PubMed]
14. Liu, N.; Mesch, M.; Weiss, T.; Hentschel, M.; Giessen, H. Infrared Perfect Absorber and Its Application As Plasmonic Sensor. *NANO Lett.* **2010**, *10*, 2342–2348. [CrossRef] [PubMed]
15. Rodrigo, D.; Limaj, O.; Janner, D.; Etezadi, D.; Javier de Abajo, F.; Pruneri, V.; Altug, H. Mid-infrared plasmonic biosensing with graphene. *Science* **2015**, *349*, 165–168. [CrossRef] [PubMed]
16. Diem, M.; Koschny, T.; Soukoulis, C.M. Wide-angle perfect absorber/thermal emitter in the terahertz regime. *Phys. Rev. B* **2009**, *79*, 033101. [CrossRef]

17. Zheng, B.Y.; Huang, T.J.; Desai, A.Y.; Wang, S.J.; Tan, L.K.; Gao, H.; Huan, A.C.H. Thermal behavior of localized surface plasmon resonance of Au/TiO$_2$ core/shell nanoparticle arrays. *Appl. Phys. Lett.* **2007**, *90*, 183117. [CrossRef]

18. Chen, A.; Cua, S.J.; Chen, P.; Chen, X.Y.; Jian, L.K. Fabrication of sub-100-nm patterns in SiO$_2$ templates by electron-beam lithography for the growth of periodic III-V semiconductor nanostructures. *Nanotechnology* **2006**, *17*, 3903–3908. [CrossRef]

19. Dolling, G.; Enkrich, C.; Wegener, M. Low-loss negative-index metamaterial at telecommunication wavelengths. *Opt. Lett.* **2006**, *31*, 1800–1802. [CrossRef]

20. Valentine, J.; Zhang, S.; Zentgraf, T.; Ulin-Avila1, E.; Genov, D.A.; Bartal, G.; Zhang, X. Three-dimensional optical metamaterial with a negative refractive index. *Nature* **2008**, *455*, 376–380. [CrossRef]

21. Zin, M.T.; Leong, K.; Wong, N.Y.; Ma, H.; Sarikawa, M.; Jen, A.K.Y. Surface-plasmon-enhanced fluorescence from periodic quantum dot arrays through distance control using biomolecular linkers. *Nanotechnology* **2009**, *20*, 015305. [CrossRef] [PubMed]

22. Cui, Y.; Xu, J.; Fung, K.H.; Jin, Y.; Kumar, A.; He, S.; Fang, N.X. A thin film broadband absorber based on multi-sized nanoantennas. *Appl. Phys. Lett.* **2011**, *99*, 253101. [CrossRef]

23. Chou, S.Y.; Krauss, P.R.; Renstrom, P.J. Imprint of sub-25 nm vias and trenches in polymers. *Science* **1996**, *272*, 85–87. [CrossRef]

24. Chou, S.Y.; Keimel, C.; Gu, J. Ultrafast and direct imprint of nanostructures in silicon. *Nature* **2002**, *417*, 835–837. [CrossRef] [PubMed]

25. Chen, L.C.; Wang, C.K.; Huang, J.B.; Hong, L.S. A nanoporous AlN layer patterned by anodic aluminum oxide and its application as a buffer layer in a GaN-based light-emitting diode. *Nanotechnology* **2009**, *20*, 085303. [CrossRef] [PubMed]

26. Yanagisita, T.; Nishio, K.; Masuda, H. Anti-Reflection structures on lenses by nanoimprinting using ordered anodic poruous alumina. *Appl. Phys. Express* **2009**, *2*, 022001. [CrossRef]

27. Reilly, T.H., III; van de Lagemaat, J.; Tenent, R.C.; Morfa, A.J.; Rowlen, K.L. Surface-plasmon enhanced transparent electrodes in organic photovoltaics. *Appl. Phys. Lett.* **2008**, *92*, 243304. [CrossRef]

28. Ren, Z.; Zhang, X.; Zhang, J.; Li, Z.; Yang, B. Building cavities in microspheres and nanospheres. *Nanotechnology* **2009**, *20*, 065305. [CrossRef]

29. Huang, C.H.; Igarashi, M.; Wone, M.; Uraoka, Y.; Fuyuki, T.; Takeguchi, M.; Yamashita, I.; Samukawa, S. Two-dimensional si-nanodisk array fabricated using bio-nano-process and neutral beam etching for realistic quantum effect devices. *Jpn. J. Appl. Phys.* **2009**, *48*, 04C187. [CrossRef]

30. Chang, C.H.; Tian, L.; Hesse, W.R.; Gal, H.; Choi, H.J.; Kim, J.G.; Siddiqui, M.; Barbastathis, G. From two-dimensional colloidal self-assembly to three-dimensional nanolithography. *Nano Lett.* **2011**, *11*, 2533–2537. [CrossRef]

31. Yoshino, M.; Ohsawa, H.; Yamanaka, A. Rapid fabrication of an ordered nano-dot array by the combination of nano plastic forming and annealing method. *J. Micromech. Microeng.* **2011**, *21*, 125017. [CrossRef]

32. Li, Z.; Yoshino, M.; Yamanaka, A. Fabrication of three dimensional ordered nanodot array structures by thermal dewetting method. *Nanotechnology* **2012**, *23*, 485303. [CrossRef] [PubMed]

33. Li, Z.; Yoshino, M.; Yamanaka, A. Regularly-formed three-dimensional gold nanodot array with controllable optical properties. *J. Micromech. Microeng.* **2014**, *24*, 045011. [CrossRef]

34. Yoshino, M.; Nagamatsu, A.; Li, Z.; Yamanaka, A.; Yamamoto, T. Optical property of metallic nanodot arrays fabricated by combination of nano plastic forming and thermal dewetting method. *Trans. JSME* **2014**, *80*, 1–13.

35. Yoshino, M.; Li, Z.; Terano, M. Theoretical and experimental study of metallic dot agglomeration induced by thermal dewetting. *ASME J. Micro Nano-Manuf.* **2015**, *3*, 021004. [CrossRef]

36. Li, Z.; Dao, T.D.; Nagao, T.; Yoshino, M. Optical properties of ordered dot-on-plate nano-sandwich arrays. *Microelectron. Eng.* **2014**, *127*, 34–39. [CrossRef]

37. Li, Z.; Dao, T.D.; Nagao, T.; Terano, M.; Yoshino, M. Fabrication of plasmonic nanopillar arrays based on nanoforming. *Microelectron. Eng.* **2015**, *139*, 7–12. [CrossRef]

38. Mullins, W.W. Theory of Thermal Grooving. *J. Appl. Phys.* **1957**, *28*, 333–339. [CrossRef]

39. Jiran, E.; Thompson, C.V. Capillary instabilities in thin films. *J. Electron. Mater.* **1990**, *19*, 1153–1160. [CrossRef]

40. McCallum, M.S.; Voorhees, P.W.; Miksis, M.J.; Davis, S.H.; Wong, H. Capillary instabilities in solid thin films: Lines. *J. Appl. Phys.* **1996**, *79*, 7604–7611. [CrossRef]

41. Danielson, D.T.; Sparacin, D.K.; Michel, J.; Kimerling, L.C. Surface-energy-driven dewetting theory of silicon-on-insulator agglomeration. *J. Appl. Phys.* **2006**, *100*, 083507. [CrossRef]
42. Young, T. An essay on the cohesion of fluids. *Philos. Trans. R. Soc. Lond.* **1805**, *95*, 65–87. [CrossRef]
43. Cheynis, F.; Bussmann, E.; Leroy, F.; Passanante, T.; Muller, P. Dewetting dynamics of silicon-on-insulator thin films. *Phys. Rev. B* **2011**, *84*, 245439. [CrossRef]
44. Rayleigh, L. On the instability of jets. *Proc. Lond. Math. Soc.* **1878**, *S1–10*, 4–13. [CrossRef]
45. Zhigal'skii, G.P.; Jones, B.K. *The Physical Properties of Thin Metal Films*; Taylor & Francis: London, UK, 2003.
46. Kuwata, H.; Tamaru, H.; Esumi, K.; Miyano, K. Resonant light scattering from metal nanoparticles: Practical analysis beyond Rayleigh approximation. *Appl. Phys. Lett.* **2003**, *83*, 4625–4627. [CrossRef]
47. Hanarp, P.; Kall, M.; Sutherland, D.S. Optical properties of short range ordered arrays of nanometer gold disks prepared by colloidal lithography. *J. Phys. Chem. B* **2003**, *107*, 5768–5772. [CrossRef]
48. Johnson, P.B.; Christy, R.W. Optical constants of the noble metals. *Phys. Rev. B* **1972**, *6*, 4370–4379. [CrossRef]
49. Bohren, C.F.; Huffman, D.R. *Absorption and Scattering of Light by Small Particles*; John Wiley & Sons: New York, NY, USA, 1983; pp. 136–148.

Article

Indium–Tin–Oxide Nanostructures for Plasmon-Enhanced Infrared Spectroscopy: A Numerical Study

Zhangbo Li [†], Zhiliang Zhang [†] and Kai Chen [*]

Institute of Photonics Technology, Jinan University, Guangzhou 511443, China; zhangboli_jnu@163.com (Z.L.); zhi_liang_zhang@163.com (Z.Z.)
* Correspondence: kaichen@jnu.edu.cn
† Equal contribution.

Received: 26 February 2019; Accepted: 3 April 2019; Published: 11 April 2019

Abstract: Plasmonic nanoantennas can significantly enhance the light–matter interactions at the nanoscale, and as a result have been used in a variety of applications such as sensing molecular vibrations in the infrared range. Indium–tin–oxide (ITO) shows metallic behavior in the infrared range, and can be used for alternative plasmonic materials. In this work, we numerically studied the optical properties of hexagonal ITO nanodisk and nanohole arrays in the mid-infrared. Field enhancement up to 10 times is observed in the simulated ITO nanostructures. Furthermore, we demonstrated the sensing of the surface phonon polariton from a 2-nm thick SiO_2 layer under the ITO disk arrays. Such periodic arrays can be readily fabricated by colloidal lithography and dry etching techniques; thus, the results shown here can help design efficient ITO nanostructures for plasmonic infrared applications.

Keywords: indium–tin oxide (ITO); plasmonics; nanoantenna; infrared spectroscopy

1. Introduction

Infrared spectroscopy is a traditional characterization technique for the accurate detection of chemical and biological species thanks to the characteristic absorption bands of the consisting chemical bonds, which form the "fingerprints" of the molecules. However, it is facing increasing difficulty when the amount of the analytes becomes less and less down to the level of monolayers or a few molecules because of the extremely small absorption cross-sections of the molecules. Surface-enhanced infrared absorption (SEIRA) spectroscopy is an enabling technique that can dramatically increase the absorption cross-sections of the molecules with significantly enhanced near-field intensities [1–7]. In general, the SEIRA substrates employ rough surfaces, metallic nanoparticles, or lithographically-defined nanostructures. Due to the lighting-rod effect and the excitation of surface plasmons, metallic nanostructures, which are usually made from Au or Ag, afford enormous near-field enhancement around the nanostructure surfaces [8–10]. The plasmon resonances of the nanoparticles can be readily tuned by controlling the shape, material, and environment of the nanoparticles [11–13]. In particular, periodic nanoparticle arrays can be achieved by e-beam lithography (EBL) or focused-ion beam (FIB), providing better control over the spectral positions of the plasmon resonances that are preferred to overlap with the molecular vibrations for enhanced sensitivity [14,15]. Besides nanoparticle-based SEIRA, there is another class of SEIRA that employs tip-based nano-IR systems and has achieved tremendous progress recently [16–18]. Such systems take advantage of the sharp-tip-induced large field enhancement to achieve ultrasensitive molecular detection with both high spectral and spatial resolution enabling promising applications for IR spectroscopy [19,20].

Traditionally, the noble metal Au is widely used as the plasmonic material due to its chemical stability, biocompatibility, and the ease of surface functionalization, which is particularly advantageous for biosensing applications. However, Au suffers from high intrinsic loss and an increasing price, which hampers its practical applications. Therefore, alternative plasmonic materials have been pursued by researchers [21–23], and light metal aluminum (Al) has been revisited as an alternative plasmonic material in the visible as well as in the infrared range [24,25]. The 2 to 3-nm thick natural oxide layer (Al_2O_3) can not only protect the underneath Al metal from oxidation, but also provide a means for stable surface functionalization [26–29]. Heavily doped oxide materials such as transparent conducting oxides (TCO) offer another option for low-loss, complementary metal-oxide-semiconductor (CMOS)-compatible plasmonic materials in the infrared range [30–35], among which indium–tin–oxide (ITO) is widely studied. ITO nanostructures have been demonstrated for refractive sensing [30] and SEIRA [32,36–38]. These ITO nanostructures are usually fabricated by top–down lithography or complex growth mechanisms; hence, more cost-effective and facile nanofabrication techniques are preferred. Colloidal lithography that employs monolayers of polymer spheres as deposition or etching masks provides an efficient method to fabricate various nanoparticle arrays [27,39,40]. In this work, we numerically studied the optical properties of ITO nanodisks and nanoholes that can be fabricated by colloidal lithography. Furthermore, we studied the coupling between the plasmon modes of the ITO nanoparticles and the surface phonon polaritons in a thin SiO_2 layer. The results shown here can serve as a guideline for future experimental studies that use ITO for infrared plasmonic sensing applications.

2. Methods

Figure 1 shows the geometrical configuration of a representative sample of the studied ITO nanodisk array. The ITO nanodisks sit on top of a Si substrate and form a hexagonal array. Such periodic structures can be fabricated using colloidal lithography and dry etching techniques at a large scale, where the polystyrene spheres act as etching masks for the underneath ITO layers. The diameter of the disks, D, can be precisely controlled by the etching time, providing a facile method of fabricating uniform large-area disk arrays. The rectangle defined by the red dashed line represents the unit cell in the finite time domain difference (FDTD) numerical simulations. Periodic boundary conditions are applied in the x and y directions, while a perfectly matched layer (PML) boundary condition is used in the z direction. A 2-nm fine mesh is applied in the regions with ITO microdisks. The incident light is polarized in the x direction and propagates in the z direction. The reflected light is collected to characterize the optical properties of the ITO disk array. The dielectric constants of Si are obtained from the handbook by Palik [41]. The dielectric constants of ITO are obtained from the Ref. 42 [42].

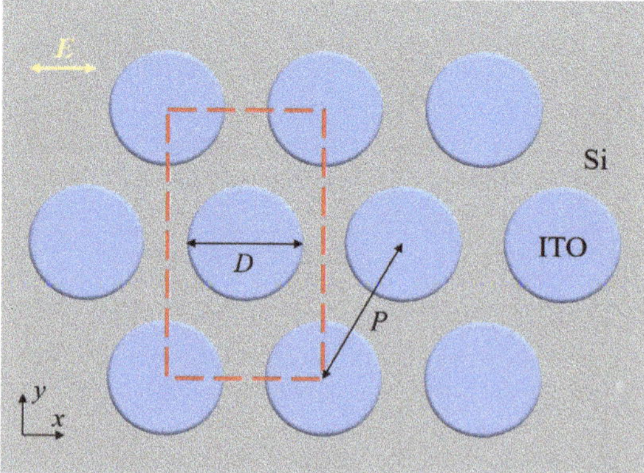

Figure 1. Illustration of the indium–tin–oxide (ITO) nanodisk array and the numerical simulation region. The ITO nanodisks with diameter D are arranged in a hexagonal array on a Si substrate. The red dashed line indicates the unit cell in the simulation, and the incident light is polarized in the x direction.

3. Results and Discussions

For disk arrays fabricated by colloidal lithography, the periodicity P of the arrays is determined by the diameter of the original polystyrene spheres used as etching masks. Thus, in our simulations, we choose the value of P according to some commercially available microspheres, namely P = 2 μm, 3 μm, and 4.4 μm, and the thickness of the disks is fixed at 100 nm. Figure 2a shows the effect of the disk diameter on the reflectance spectra of the samples. As the most popular TCO material, ITO displays metallic behavior in the infrared wavelength range. The real part (ε_1) of the dielectric constants of the ITO (obtained from Ref. 42) crosses zero at 1.6 μm from the positive to the negative region. The reflectance spectra of the disk arrays show similar trends to those of metallic disk arrays. As the disk diameter D increases, the resonance peak shifts to the longer wavelength range, and the peak intensity increases at the same time. Figure 2b,c show the near-field distribution of the electric field at 8.24 μm for the disk array with D = 1 μm and P = 2 μm. Clearly, the near field shows enhancement around the disk edges at resonance wavelengths, indicating the dipolar character of the plasmon resonance. As shown in Figure 2b, the hot spots are located at the ITO/Si interface.

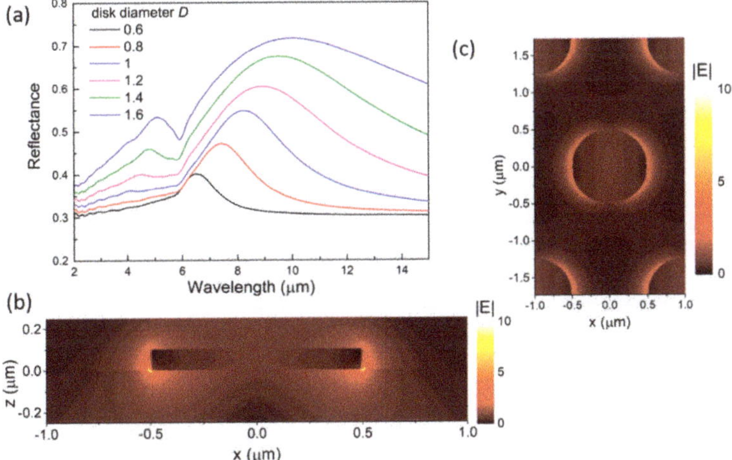

Figure 2. Optical properties of an ITO disk array with P = 2 μm. (**a**) The reflectance spectra of the ITO disks with different diameters. (**b**) The electric near-field profile inside the xz cross-section at 8.24 μm for the disk with D = 1 μm. The enhanced field is concentrated near the disk edges and close to the Si substrate. (**c**) The electric near-field profile of the disk array as in (**b**) at 8.24 μm. The magnitude of the electric field is plotted in panels (**b**) and (**c**).

Figure 3 shows the effect of the periodicity of the array on the reflectance spectra. The diameter of the disks is fixed at 1 μm. With large periodicity, the filling factor of the disk arrays is low, and thus a broad and shallow peak is observed. As the periodicity becomes smaller, the disk density becomes larger, and more light is reflected and scattered. It is noted that lattice coupling exists inside a nanoparticle array, and the coupling strength depends on the array geometry, particularly the period. At a certain period, constructive interference occurs, leading to a much narrower resonance peak [2]. In our case, such constructive interference happens when the period is ~3 μm, and therefore, smaller or larger periods result in broader resonance peaks.

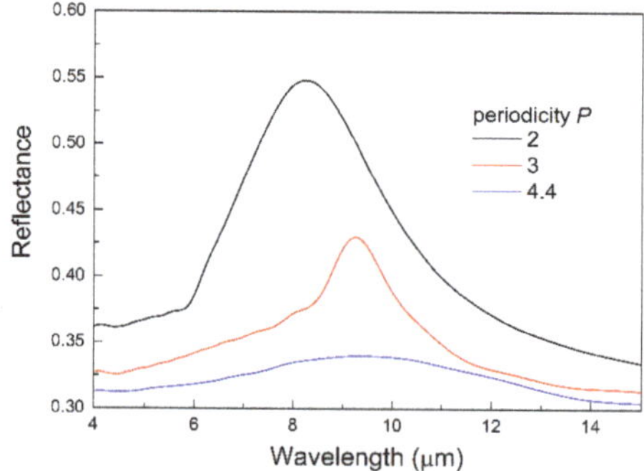

Figure 3. The effect of the periodicity P on the reflectance spectra of the disk arrays with D = 1 μm.

The complementary nanostructures to the disk arrays, i.e., the hexagonal nanohole arrays, can also be easily fabricated by colloidal lithography, where size-reduced spheres can be used as deposition masks for ITO. Therefore, we also numerically studied ITO nanohole arrays, and the results are shown in Figure 4. Similar to the disk arrays in Figure 2, the periodicity of the nanohole arrays is fixed at 2 μm, and the thickness is fixed at 200 nm. We only vary the diameters of the nanoholes. Nanohole arrays are known to support extraordinary optical transmission (EOT) due to the excitation of the localized surface plasmon modes of the nanoholes. A major EOT peak is observed in the transmittance spectra at ~8 μm. As the hole diameter increases, the peak intensity increases while the peak position stays in the vicinity of 8 μm in contrast to the large red-shift observed in the ITO disk arrays, as shown in Figure 2a. The different optical behavior can be attributed to the complex mechanism behind the EOT phenomenon where the transmission is mediated by the localized plasmon mode of individual nanoholes and the collective optical modes of the arrays [43,44]. Figure 4b displays the near-field distribution of the electric field at the resonance peak (8 μm).

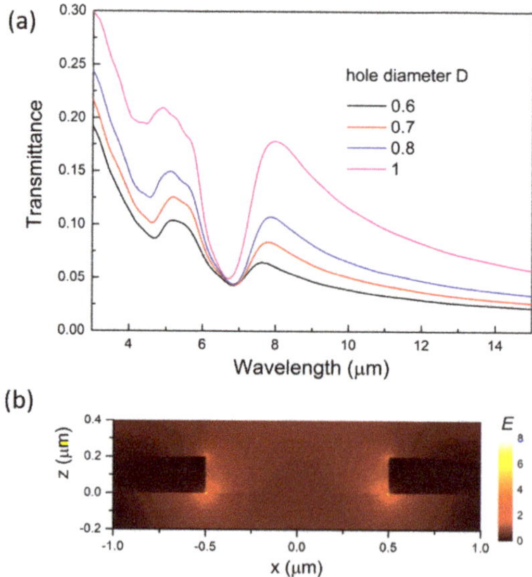

Figure 4. The optical properties of ITO nanohole arrays. (**a**) The reflectance spectra of the ITO nanohole arrays with different hole diameters. The periodicity of the arrays is 2 μm. (**b**) The electric near-field distribution of the nanohole at 8 μm. The diameter of the nanohole is 1 μm, and the color bar shows the magnitude of the electric field.

It is noted that the free carrier density of ITO depends on the preparation techniques as well as post-deposition annealing [42,45]. Thus, the dielectric constant of ITO can be readily tuned via various parameters in deposition and annealing, providing an effective means of controlling the optical properties of ITO nanoantennas and making ITO an attractive material for infrared plasmonics [30]. The results presented here are based on the dielectric constants in earlier reports [42].

The infrared nanoantennas can concentrate the electromagnetic field into subwavelength volume, leading to significantly enhanced near-field intensity, i.e., hot spots, which have been utilized in various SEIRA reports with different nanoantenna designs and materials, including TCO materials [46]. As we shown in Figure 2, such near-field enhancement occurs around the ITO disks. We further investigated their sensing capability by placing a 2-nm thick SiO_2 layer under the ITO disks and detecting the surface phonon polariton (SPhP) signal from the SiO_2. The collected reflectance spectra are shown in Figure 5a, where the simulated disk arrays have the same dimensions as those in Figure 2a. Compared with Figure 2a, distinct dips are observed on these spectra, and the dips appear around 8 μm, which is inside the range of the Fuchs–Kliewer surface phonon polariton of SiO_2, and is consistent with previous reports [40]. It is noted that the presence of ITO microdisks facilitate the excitation of the SiO_2 SPhP, as the required additional momentum could be provided by the scattered light from the microdisks [47]. The shape and intensity of the dip are apparently affected by the detuning between the vibrations and the plasmon resonances. When the plasmon peak shows a good overlap with the vibration, as for ITO disks with D = 1 μm, the intensity of the dip reaches the maximum. Thus, we chose this ITO disk array (P = 2 μm and D = 1 μm) to study the effect of the SiO_2 layer thickness, as shown in Figure 5b. As the SiO_2 layer thickness increases, the intensity of the dip increases, because more SiO_2 is included inside the volume of the hot spots, resulting in a stronger coupling between the plasmon mode and the vibration.

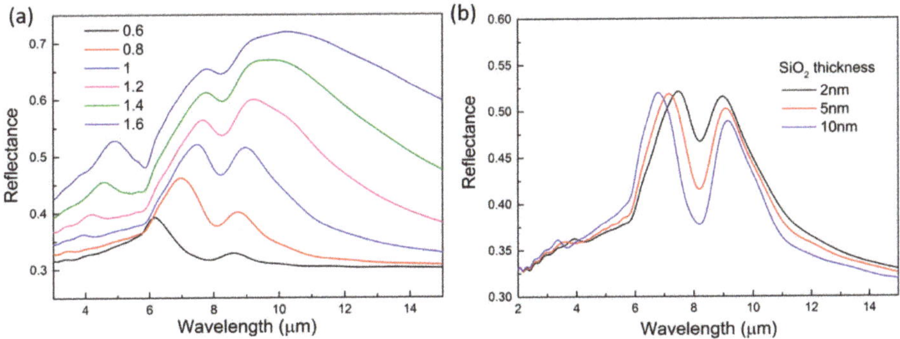

Figure 5. Sensing of the surface phonon polariton from SiO_2 with ITO nanodisks. (**a**) Reflectance spectra of the ITO disk arrays with a 2-nm thick SiO_2 layer underneath the arrays. The diameter of the disks is changed to tune the plasmon resonances and study the coupling between the plasmon mode and the SiO_2 vibration. (**b**) Reflectance spectra of the disk array (P = 2 μm and D = 1 μm) with different SiO_2 layer thickness.

4. Conclusions

In summary, we have numerically studied the optical properties of hexagonal ITO nanodisk arrays and nanohole arrays. We specifically tune the dimensions of the nanostructures to shift their plasmon resonances in the infrared wavelength range, making them suitable for SEIRA sensing applications. Both the ITO disks and nanoholes show distinct plasmon resonances in the mid-infrared range, and field enhancement up to 10 times is observed in the ITO nanostructures. The resonances of the ITO disk arrays can be readily tuned in a broad wavelength range. We also demonstrate the excitation and sensing of the surface phonon polariton signal from a 2-nm thick SiO_2 layer under the disk arrays. Both the ITO nanodisk and nanohole arrays can be fabricated on a large scale using colloidal lithography and dry etching methods. Thus, the results shown here can be useful for the design and application of ITO nanostructures operating in the mid-infrared wavelength range.

Author Contributions: Conceptualization, K.C.; Simulations, Z.L. and Z.Z.; Data Analysis, Z.L., Z.Z. and K.C.; Writing—Original Draft Preparation, K.C.; Writing—Review & Editing, Z.L., Z.Z. and K.C.; Supervision and Project Administration, K.C.

Funding: This research was funded by Guangdong Provincial Innovation and Entrepreneurship Project (grant no. 2016ZT06D081).

Acknowledgments: The authors acknowledge the financial support from Guangdong Provincial Innovation and Entrepreneurship Project (grant no. 2016ZT06D081) and the startup funding from Jinan University.

Conflicts of Interest: The authors declare no competing financial interest.

References

1. Neubrech, F.; Pucci, A.; Cornelius, T.W.; Karim, S.; Garcia-Etxarri, A.; Aizpurua, J. Resonant Plasmonic and Vibrational Coupling in a Tailored Nanoantenna for Infrared Detection. *Phys. Rev. Lett.* **2008**, *101*, 157403. [CrossRef]

2. Adato, R.; Yanik, A.A.; Amsden, J.J.; Kaplan, D.L.; Omenetto, F.G.; Hong, M.K.; Erramilli, S.; Altug, H. Ultra-sensitive vibrational spectroscopy of protein monolayers with plasmonic nanoantenna arrays. *Proc. Natl. Acad. Sci. USA* **2009**, *106*, 19227–19232. [CrossRef]

3. Wu, C.; Khanikaev, A.B.; Adato, R.; Arju, N.; Yanik, A.A.; Altug, H.; Shvets, G. Fano-resonant asymmetric metamaterials for ultrasensitive spectroscopy and identification of molecular monolayers. *Nat. Mater.* **2012**, *11*, 69–75. [CrossRef] [PubMed]

4. Chen, K.; Adato, R.; Altug, H. Dual-Band Perfect Absorber for Multispectral Plasmon-Enhanced Infrared Spectroscopy. *ACS Nano* **2012**, *6*, 7998–8006. [CrossRef] [PubMed]

5. Hoang, C.V.; Oyama, M.; Saito, O.; Aono, M.; Nagao, T. Monitoring the Presence of Ionic Mercury in Environmental Water by Plasmon-Enhanced Infrared Spectroscopy. *Sci. Rep.* **2013**, *3*, 1175. [CrossRef] [PubMed]

6. Cetin, A.E.; Etezadi, D.; Altug, H. Accessible Nearfields by Nanoantennas on Nanopedestals for Ultrasensitive Vibrational Spectroscopy. *Adv. Opt. Mater.* **2014**, *2*, 866–872. [CrossRef]

7. Bagheri, S.; Weber, K.; Gissibl, T.; Weiss, T.; Neubrech, F.; Giessen, H. Fabrication of Square-Centimeter Plasmonic Nanoantenna Arrays by Femtosecond Direct Laser Writing Lithography: Effects of Collective Excitations on SEIRA Enhancement. *ACS Photonics* **2015**, *2*, 779–786. [CrossRef]

8. Neubrech, F.; Weber, D.; Katzmann, J.; Huck, C.; Toma, A.; Di Fabrizio, E.; Pucci, A.; Härtling, T. Infrared Optical Properties of Nanoantenna Dimers with Photochemically Narrowed Gaps in the 5 nm Regime. *ACS Nano* **2012**, *6*, 7326–7332. [CrossRef]

9. Hoffmann, J.M.; Janssen, H.; Chigrin, D.N.; Taubner, T. Enhanced infrared spectroscopy using small-gap antennas prepared with two-step evaporation nanosphere lithography. *Opt. Express* **2014**, *22*, 14425–14432. [CrossRef] [PubMed]

10. Enders, D.; Nagao, T.; Pucci, A.; Nakayama, T.; Aono, M. Surface-enhanced ATR-IR spectroscopy with interface-grown plasmonic gold-island films near the percolation threshold. *Phys. Chem. Chem. Phys.* **2011**, *13*, 4935–4941. [CrossRef] [PubMed]

11. Gaspar, D.; Pimentel, A.C.; Mateus, T.; Leitão, J.P.; Soares, J.; Falcão, B.P.; Araújo, A.; Vicente, A.; Filonovich, S.A.; Águas, H.; et al. Influence of the layer thickness in plasmonic gold nanoparticles produced by thermal evaporation. *Sci. Rep.* **2013**, *3*, 1469. [CrossRef] [PubMed]

12. Chen, K.; Leong Eunice Sok, P.; Rukavina, M.; Nagao, T.; Liu Yan, J.; Zheng, Y. Active molecular plasmonics: Tuning surface plasmon resonances by exploiting molecular dimensions. *Nanophotonics* **2015**, *4*, 186–197. [CrossRef]

13. Araújo, A.; Mendes, M.J.; Mateus, T.; Vicente, A.; Nunes, D.; Calmeiro, T.; Fortunato, E.; Águas, H.; Martins, R. Influence of the Substrate on the Morphology of Self-Assembled Silver Nanoparticles by Rapid Thermal Annealing. *J. Phys. Chem. C* **2016**, *120*, 18235–18242. [CrossRef]

14. Maß, T.W.W.; Taubner, T. Incident Angle-Tuning of Infrared Antenna Array Resonances for Molecular Sensing. *ACS Photonics* **2015**, *2*, 1498–1504. [CrossRef]

15. Weber, D.; Albella, P.; Alonso-González, P.; Neubrech, F.; Gui, H.; Nagao, T.; Hillenbrand, R.; Aizpurua, J.; Pucci, A. Longitudinal and transverse coupling in infrared gold nanoantenna arrays: Long range versus short range interaction regimes. *Opt. Express* **2011**, *19*, 15047–15061. [CrossRef]

16. Chen, J.; Badioli, M.; Alonso-Gonzalez, P.; Thongrattanasiri, S.; Huth, F.; Osmond, J.; Spasenovic, M.; Centeno, A.; Pesquera, A.; Godignon, P.; et al. Optical nano-imaging of gate-tunable graphene plasmons. *Nature* **2012**, *487*, 77–81. [CrossRef] [PubMed]

17. Alonso-González, P.; Albella, P.; Schnell, M.; Chen, J.; Huth, F.; García-Etxarri, A.; Casanova, F.; Golmar, F.; Arzubiaga, L.; Hueso, L.E.; et al. Resolving the electromagnetic mechanism of surface-enhanced light scattering at single hot spots. *Nat. Commun.* **2012**, *3*, 684. [CrossRef] [PubMed]

18. Taubner, T.; Eilmann, F.; Hillenbrand, R. Nanoscale-resolved subsurface imaging by scattering-type near-field optical microscopy. *Opt. Express* **2005**, *13*, 8893–8899. [CrossRef]

19. Lewin, M.; Baeumer, C.; Gunkel, F.; Schwedt, A.; Gaussmann, F.; Wueppen, J.; Meuffels, P.; Jungbluth, B.; Mayer, J.; Dittmann, R.; et al. Nanospectroscopy of Infrared Phonon Resonance Enables Local Quantification of Electronic Properties in Doped $SrTiO_3$ Ceramics. *Adv. Funct. Mater.* **2018**, *28*, 1802834. [CrossRef]

20. Xu, X.G.; Rang, M.; Craig, I.M.; Raschke, M.B. Pushing the Sample-Size Limit of Infrared Vibrational Nanospectroscopy: From Monolayer toward Single Molecule Sensitivity. *J. Phys. Chem. Lett.* **2012**, *3*, 1836–1841. [CrossRef]

21. West, P.R.; Ishii, S.; Naik, G.V.; Emani, N.K.; Shalaev, V.M.; Boltasseva, A. Searching for better plasmonic materials. *Laser Photonics Rev.* **2010**, *4*, 795–808. [CrossRef]

22. Kumar, M.; Umezawa, N.; Ishii, S.; Nagao, T. Examining the Performance of Refractory Conductive Ceramics as Plasmonic Materials: A Theoretical Approach. *ACS Photonics* **2016**, *3*, 43–50. [CrossRef]

23. Yang, X.; Sun, Z.; Low, T.; Hu, H.; Guo, X.; García de Abajo, F.J.; Avouris, P.; Dai, Q. Nanomaterial-Based Plasmon-Enhanced Infrared Spectroscopy. *Adv. Mater.* **2018**, *30*, 1704896. [CrossRef] [PubMed]

24. Knight, M.W.; King, N.S.; Liu, L.; Everitt, H.O.; Nordlander, P.; Halas, N.J. Aluminum for Plasmonics. *ACS Nano* **2013**, *8*, 834–840. [CrossRef] [PubMed]

25. Lecarme, O.; Sun, Q.; Ueno, K.; Misawa, H. Robust and Versatile Light Absorption at Near-Infrared Wavelengths by Plasmonic Aluminum Nanorods. *ACS Photonics* **2014**, *1*, 538–546. [CrossRef]

26. Cerjan, B.; Yang, X.; Nordlander, P.; Halas, N.J. Asymmetric Aluminum Antennas for Self-Calibrating Surface-Enhanced Infrared Absorption Spectroscopy. *ACS Photonics* **2016**, *3*, 354–360. [CrossRef]

27. Chen, K.; Dao, T.D.; Ishii, S.; Aono, M.; Nagao, T. Infrared Aluminum Metamaterial Perfect Absorbers for Plasmon-Enhanced Infrared Spectroscopy. *Adv. Funct. Mater.* **2015**, *25*, 6637–6643. [CrossRef]

28. Ayas, S.; Topal, A.E.; Cupallari, A.; Güner, H.; Bakan, G.; Dana, A. Exploiting Native Al_2O_3 for Multispectral Aluminum Plasmonics. *ACS Photonics* **2014**, *1*, 1313–1321. [CrossRef]

29. Canalejas-Tejero, V.; Herranz, S.; Bellingham, A.; Moreno-Bondi, M.C.; Barrios, C.A. Passivated aluminum nanohole arrays for label-free biosensing applications. *ACS Appl. Mater. Interfaces* **2014**, *6*, 1005–1010. [CrossRef] [PubMed]

30. Li, S.Q.; Guo, P.; Zhang, L.; Zhou, W.; Odom, T.W.; Seideman, T.; Ketterson, J.B.; Chang, R.P.H. Infrared Plasmonics with Indium–Tin-Oxide Nanorod Arrays. *ACS Nano* **2011**, *5*, 9161–9170. [CrossRef] [PubMed]

31. Babicheva, V.E.; Kinsey, N.; Naik, G.V.; Ferrera, M.; Lavrinenko, A.V.; Shalaev, V.M.; Boltasseva, A. Towards CMOS-compatible nanophotonics: Ultra-compact modulators using alternative plasmonic materials. *Opt. Express* **2013**, *21*, 27326–27337. [CrossRef] [PubMed]

32. Abb, M.; Wang, Y.; Papasimakis, N.; de Groot, C.H.; Muskens, O.L. Surface-Enhanced Infrared Spectroscopy Using Metal Oxide Plasmonic Antenna Arrays. *Nano Lett.* **2014**, *14*, 346–352. [CrossRef] [PubMed]

33. Kim, J.; Dutta, A.; Memarzadeh, B.; Kildishev, A.V.; Mosallaei, H.; Boltasseva, A. Zinc Oxide Based Plasmonic Multilayer Resonator: Localized and Gap Surface Plasmon in the Infrared. *ACS Photonics* **2015**, *2*, 1224–1230. [CrossRef]

34. Guo, P.; Schaller, R.D.; Ketterson, J.B.; Chang, R.P.H. Ultrafast switching of tunable infrared plasmons in indium tin oxide nanorod arrays with large absolute amplitude. *Nat. Photonics* **2016**, *10*, 267–273. [CrossRef]

35. Wang, Y.; Overvig, A.C.; Shrestha, S.; Zhang, R.; Wang, R.; Yu, N.; Dal Negro, L. Tunability of indium tin oxide materials for mid-infrared plasmonics applications. *Opt. Mater. Express* **2017**, *7*, 2727–2739. [CrossRef]

36. Chen, K.; Guo, P.; Dao, T.D.; Li, S.-Q.; Ishiii, S.; Nagao, T.; Chang, R.P.H. Protein-Functionalized Indium-Tin Oxide Nanoantenna Arrays for Selective Infrared Biosensing. *Adv. Opt. Mater.* **2017**, *5*, 1700091. [CrossRef]

37. Kamakura, R.; Takeishi, T.; Murai, S.; Fujita, K.; Tanaka, K. Surface-Enhanced Infrared Absorption for the Periodic Array of Indium Tin Oxide and Gold Microdiscs: Effect of in-Plane Light Diffraction. *ACS Photonics* **2018**, *5*, 2602–2608. [CrossRef]

38. D'apuzzo, F.; Esposito, M.; Cuscunà, M.; Cannavale, A.; Gambino, S.; Lio, G.E.; De Luca, A.; Gigli, G.; Lupi, S. Mid-Infrared Plasmonic Excitation in Indium Tin Oxide Microhole Arrays. *ACS Photonics* **2018**, *5*, 2431–2436. [CrossRef]

39. Dao, T.D.; Chen, K.; Ishii, S.; Ohi, A.; Nabatame, T.; Kitajima, M.; Nagao, T. Infrared Perfect Absorbers Fabricated by Colloidal Mask Etching of Al–Al$_2$O$_3$–Al Trilayers. *ACS Photonics* **2015**, *2*, 964–970. [CrossRef]

40. Chen, K.; Duy Dao, T.; Nagao, T. Tunable Nanoantennas for Surface Enhanced Infrared Absorption Spectroscopy by Colloidal Lithography and Post-Fabrication Etching. *Sci. Rep.* **2017**, *7*, 44069. [CrossRef] [PubMed]

41. Palik, E.D. *Handbook of Optical Constants of Solids*, 3rd ed.; Academic Press: New York, NY, USA, 1998.

42. Tamanai, A.; Dao, T.D.; Sendner, M.; Nagao, T.; Pucci, A. Mid-infrared optical and electrical properties of indium tin oxide films. *Phys. Status Solidi A* **2017**, *214*, 1600467. [CrossRef]

43. Garcia-Vidal, F.J.; Martin-Moreno, L.; Ebbesen, T.W.; Kuipers, L. Light passing through subwavelength apertures. *Rev. Mod. Phys.* **2010**, *82*, 729–787. [CrossRef]

44. Laux, E.; Genet, C.; Ebbesen, T.W. Enhanced optical transmission at the cutoff transition. *Opt. Express* **2009**, *17*, 6920–6930. [CrossRef] [PubMed]

45. Baía, I.; Quintela, M.; Mendes, L.; Nunes, P.; Martins, R. Performances exhibited by large area ITO layers produced by r.f. magnetron sputtering. *Thin Solid Films* **1999**, *337*, 171–175. [CrossRef]

46. Neubrech, F.; Huck, C.; Weber, K.; Pucci, A.; Giessen, H. Surface-Enhanced Infrared Spectroscopy Using Resonant Nanoantennas. *Chem. Rev.* **2017**, *117*, 5110–5145. [CrossRef] [PubMed]

47. Neubrech, F.; Weber, D.; Enders, D.; Nagao, T.; Pucci, A. Antenna Sensing of Surface Phonon Polaritons. *J. Phys. Chem. C* **2010**, *114*, 7299–7301. [CrossRef]

Article

Effect of Etching Depth on Threshold Characteristics of GaSb-Based Middle Infrared Photonic-Crystal Surface-Emitting Lasers

Zong-Lin Li [1], Shen-Chieh Lin [1], Gray Lin [1,*], Hui-Wen Cheng [1,2], Kien-Wen Sun [3] and Chien-Ping Lee [1]

[1] Department of Electronics Engineering and Institute of Electronics, National Chiao Tung University, Hsinchu City 30010, Taiwan; martin323261@gmail.com (Z.-L.L.); z810481@gmail.com (S.-C.L.); hwcheng2011@gmail.com (H.-W.C.); cplee@mail.nctu.edu.tw (C.-P.L.)
[2] Center for Nano Science and Technology, National Chiao Tung University, Hsinchu City 30010, Taiwan
[3] Department of Applied Chemistry, National Chiao Tung University, Hsinchu City 30010, Taiwan; kwsun@mail.nctu.edu.tw
* Correspondence: graylin@mail.nctu.edu.tw; Tel.: +886-3-513-1289

Received: 28 February 2019; Accepted: 13 March 2019; Published: 14 March 2019

Abstract: We study the effect of etching depth on the threshold characteristics of GaSb-based middle infrared (Mid-IR) photonic-crystal surface-emitting lasers (PCSELs) with different lattice periods. The below-threshold emission spectra are measured to identify the bandgap as well as band-edge modes. Moreover, the bandgap separation widens with increasing etching depth as a result of enhanced diffraction feedback coupling. However, the coupling is nearly independent of lattice period. The relationship between threshold gain and Bragg detuning is also experimentally determined for PCSELs and is similar to that calculated theoretically for one-dimensional distributed feedback lasers.

Keywords: photonic crystals; surface-emitting lasers; middle infrared lasers; GaSb-based lasers

---

1. Introduction

Semiconductor lasers emitting in the middle infrared (Mid-IR) range have promising applications in gas sensing, environmental monitoring, and military explosives detection [1–3]. Gallium antimonide (GaSb) and GaSb-related semiconductors are ideally suited for such light emitters because of the narrow energy bandgap as well as the Type-I quantum well (QW) heterostructure [2]. It is worth mentioning that Mid-IR sensors based on tunable diode laser absorption spectroscopy (TDLAS) render the lasers in single spectral mode with a narrow linewidth.

In recent years, photonic-crystal surface-emitting lasers (PCSELs) have attracted a lot of attention because of their narrow spectral linewidth, high output power, and small beam divergence angle [4–6]. By properly designing two-dimensional (2D) photonic crystals (PhC) that satisfy a specific Bragg condition, light waves from gain media can couple with PhC, whereby a 2D cavity mode is constructed to produce lasing emissions from the surface of the device. GaSb-based PCSELs are designed in connection to the abovementioned TDLAS sensors; however, only optically pumped devices are successfully demonstrated [5,7], while electrically pumped ones are still in development [8].

The design and fabrication of PCSELs without a complicated technology of regrowth or fusion bonding is preferred, but weak diffraction feedback coupling between the PhC layer and QW active region is resulted because the optical mode is pushed away from the PhC layer by low-index ambient and etched holes. To compensate for weak coupling, the etching depth of PhC holes is increased to reduce the threshold pumping density. Prior works involved two or three etching depths [9,10]; no systematic study on depth effect has been conducted.

In this work, we report the effect of etching depth on the threshold characteristics of PCSELs with different lattice periods. GaSb-based gain media are selected as they are easily pumped with a low threshold pumping density [5,7]. The below-threshold emission spectra are also measured to identify the bandgap as well as the band-edge modes. The bandgap separation is analyzed with respect to etching depth and lattice period. Finally, the relationship between threshold gain and Bragg detuning is experimentally determined and discussed.

2. Materials and Methods

The sample in this work was grown on (001) n-type GaSb substrates using a Veeco GEN II molecular beam epitaxy system. It consisted of, from the bottom upward, a 200-nm GaSb buffer layer, a 2000-nm $Al_{0.85}Ga_{0.15}As_{0.07}Sb_{0.93}$ bottom-cladding layer, a 150-nm $Al_{0.3}Ga_{0.7}As_{0.02}Sb_{0.98}$ separate confinement layer (SCL), two layers of 10-nm $In_{0.35}Ga_{0.65}As_{0.14}Sb_{0.86}$ quantum well (QW) spaced by 20-nm $Al_{0.3}Ga_{0.7}As_{0.02}Sb_{0.98}$ layer, a 200-nm $Al_{0.3}Ga_{0.7}As_{0.02}Sb_{0.98}$ SCL, a 200-nm top-cladding layer $Al_{0.5}Ga_{0.5}As_{0.04}Sb_{0.96}$, and a 200-nm GaSb capping layer. The entire structure is shown in Figure 1a.

A 150-nm-thick Si_3N_4 layer was first deposited as a hard mask by plasma-enhanced chemical vapor deposition. The PhC structure was then fabricated using electron-beam lithography and an inductively coupled plasma reactive ion etching system. Finally, the hard mask was removed before optical pumping. The pattern of PhC was designed in a square lattice with circularly shaped air holes. An array of 5-by-5 PhC regions (five rows indexed 1 to 5 by five columns indexed A to E) was patterned in a small wafer and cut into five pieces for different etching depth. The lattice periods varied from 620 nm to 640 nm in steps of 5 nm and were labeled 1 to 5 with increasing period. PhC holes with different depths of 150, 200, 250, 300, and 350 nm were dry-etched and labeled A to E with increasing depth. Therefore, a total of 25 PCSEL devices were prepared for the measurement, i.e., devices A1 to A5, B1 to B5, C1 to C5, D1 to D5, and E1 to E5. The filling factor, defined as the ratio of hole area within unit lattice, was 0.1 and each device area was 300×300 μm^2. Figure 1b is the cross-sectional scanning electron microscope (SEM) image of a PCSEL.

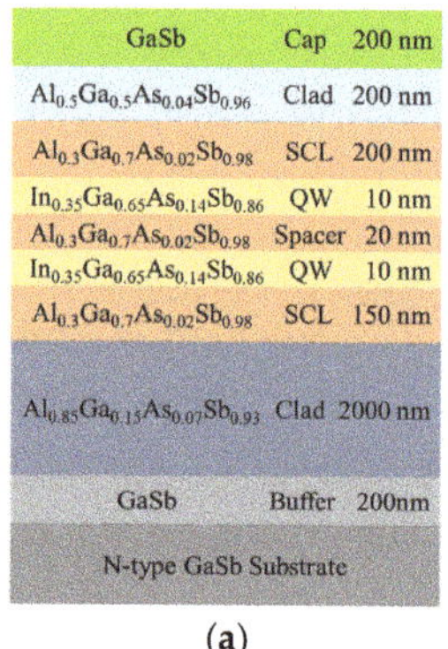

(a)

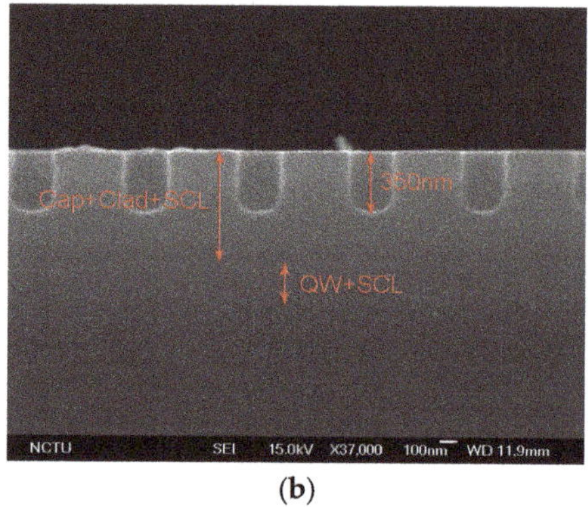

(b)

Figure 1. (a) The schematic diagram of sample structure. (b) The cross-sectional SEM images of a PCSEL.

The PCSEL devices were fixed on a copper plate. The temperature of the plate was controlled at 20 °C using a temperature controller (LDT-5545B, ILX Lightwave Co., Bozeman, MT, USA). The devices were pumped by a 1064-nm pulsed fiber laser (MOPA-PS/CDRH 12V, Multiwave Photonics S.A., Maia, Portugal) with a pulse width of 100 ns and a repetition rate of 5 kHz. It should be noted that a CaF_2 lens (LA5370, Thorlabs Inc., Newton, NJ, USA,) was used to focus the pumping laser beam as well as to collect the Mid-IR emissions normally to the device surface. The illuminated spot was 200 μm in diameter, which approximately filled the entire area of the device. A longpass filter was used to block the 1064 nm pumping light. Finally, another CaF_2 lens was used to focus the Mid-IR light into a monochromator (iHR 320, Horiba Jobin Yvon Inc., USA) with a thermoelectrically cooled InGaAsSb detector (IGA2.2-010-TC, Electro-Optical Systems Inc., Phoenixville, PA, USA). The resolution of the monochromator was 0.1 nm. Figure 2 shows the setup of the optical pumping system.

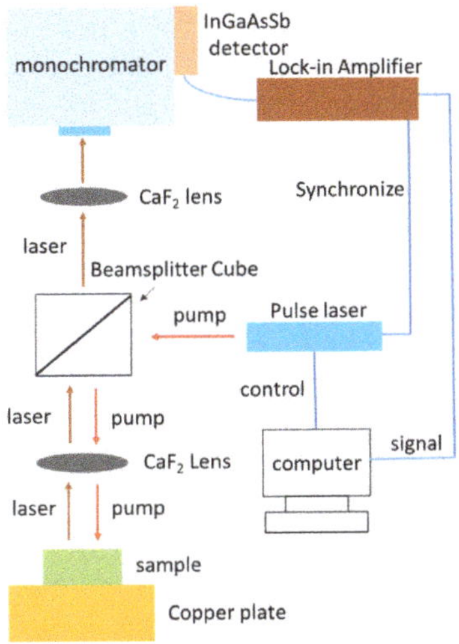

Figure 2. The measurement setup of optical pumping system.

3. Results and Discussions

The 25 devices were optically pumped to exhibit surface lasing emissions. The lasing spectra were acquired at pumping power of 1.2 times the threshold. Figure 3a shows the normalized spectra for devices A4, B4, C4, D4, and E4 (with a fixed lattice period of 635 nm). The spectral full width at half maximum (FWHM) ranges from 0.2 nm to 0.3 nm, and the corresponding quality factor (Q-factor) distributes from a little more than 7000 to over 10,000. The peak lasing wavelengths of the 25 devices are plotted against individual etching depths in Figure 3b. The linear dependence is fitted for five different lattice periods with an almost equal slope of -0.05, i.e., a wavelength shift of -5 nm for every 100-nm increase in etching depth. At a fixed etching depth, the wavelength shift is about $+3.14$ nm per nm increase in lattice period, which is consistent with InP- and GaAs-based PCSELs [10,11].

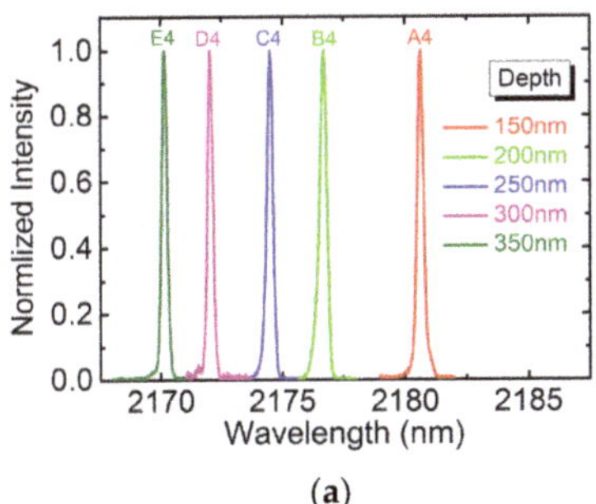

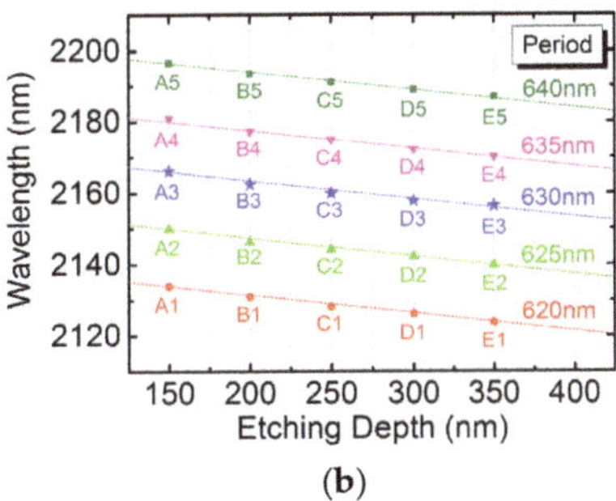

Figure 3. (a) The normalized lasing spectrum for devices with fixed lattice period of 635 nm. (b) The peak lasing wavelength versus etching depth for 25 devices.

The RT photo-luminescence (PL) was peaked around 2172 nm with spectral width at 95% intensity of about 50 nm (not shown). The 635-nm-period devices (A4 to E4), which lased around the gain peak, exhibited the lowest pumping threshold among the different devices. Figure 4a shows the curves of light-in versus light-out (*L-L*) for devices A4 to E4. The slope efficiency is not discussed because there may be large deviations between measurements. The plot of threshold power density versus etching depth is shown in Figure 4b. The deeper the PhC holes are etched, the more enhanced feedback couplings between PhC layers and QW active region are achieved in which lower threshold gain is resulted. Therefore, the threshold power decreases rapidly with increasing depth and then levels off to become saturated. The dependence of threshold power density on etching depth can be fitted by an exponential function with an offset of 225. The offset is attributed to minimum achievable threshold gain, which will be discussed later.

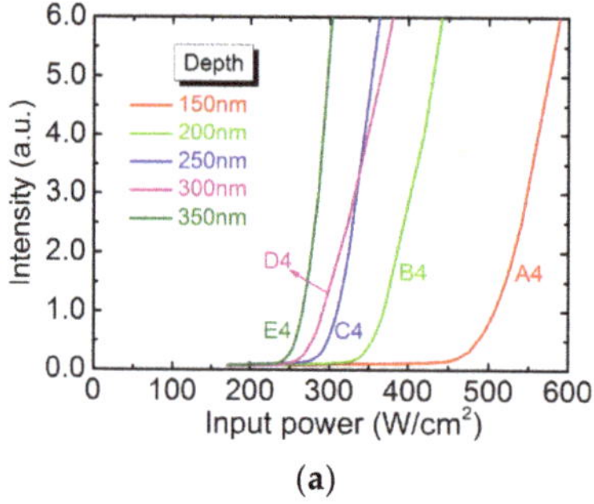
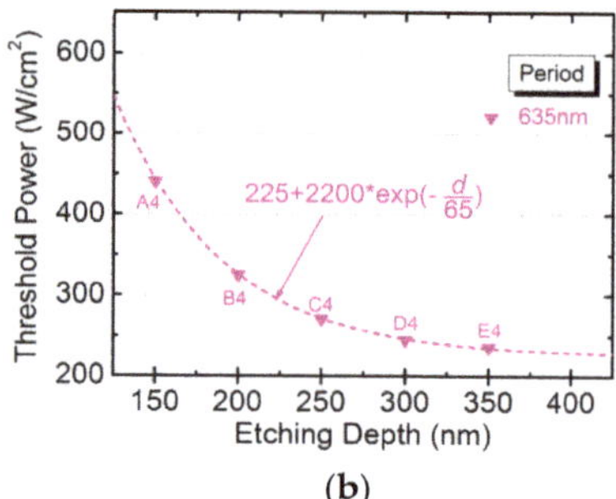

Figure 4. (a) The light-in versus light-out (*L-L*) curves and (b) the dependence of threshold power density on the etching depth for 635-nm-period devices.

Figure 5a shows the etching depth dependence of threshold power density for 620- and 630-nm-period devices (A1 to E1 and A3 to E3). They are fitted by exponential functions with the offsets of 315 and 285 for 620- and 630-nm-period devices, respectively. Based on our previous works [9–11], devices with less gain-cavity offset and deeper etching depth exhibited a lower threshold pumping density. Therefore, we excluded devices D1 and E3 from fitting according to the above criteria. The threshold power densities for device A2 to E2 and A5 to E5 are also plotted again etching depths in Figure 5b. However, no exponential dependence is observed for the 625- and 640-nm-period devices. Note that the gain-cavity offsets fell between 620- and 635-nm-period devices, but higher pumping thresholds were observed for devices B2, B5, C2, C5, and E2. We attributed the causes to growth imperfection, process variation, and/or a misalignment in focusing. As a result, the threshold pumping densities of the 625- and 640-nm-period devices are not analyzed afterwards.

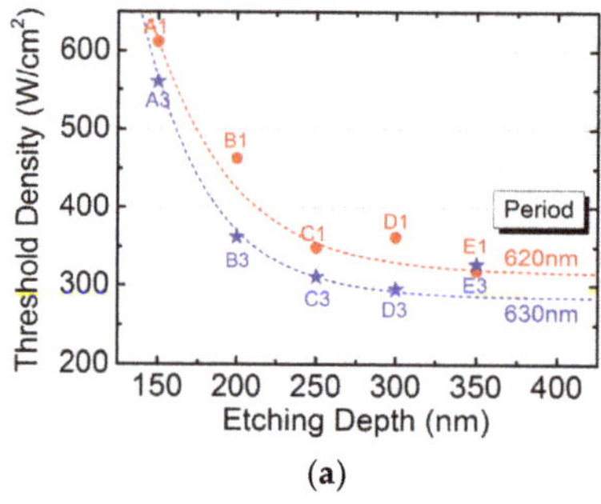
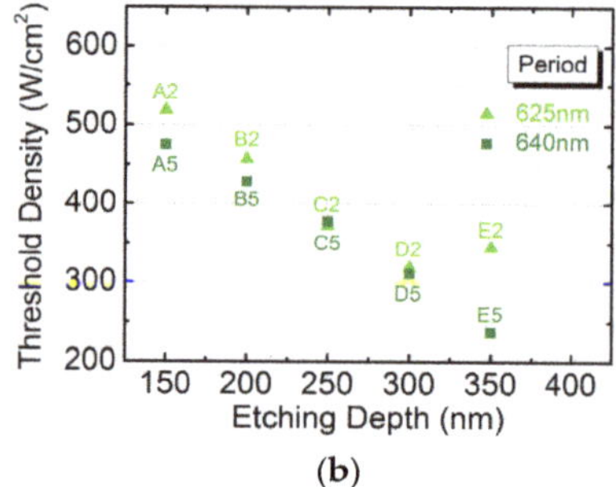

Figure 5. The dependence of threshold power density on the etching depth for (a) 620- and 630-nm-period devices (A1 to E1 and A3 to E3) as well as (b) 625- and 640-nm-period devices (A2 to E2 and A5 to E5).

Because feedback coupling between PhC layers and QW active region is enhanced with increasing etching depth, the normalized frequency difference between band-edge modes is expected to widen. Based on the measurement setup in Figure 2, we collected below-threshold emissions by optical lens with numerical aperture of about 0.3. Since angular information cannot be resolved, only band-edge modes B and C are identified by their intensity contrast with mode bandgap. Figure 6a shows the above-mentioned spectra for 635-nm-period devices (A4 to E4). The longer-wavelength and shorter-wavelength peaks in the immediate vicinity of intensity bandgap are ascribed to band-edge modes B and C, respectively. Moreover, the bandgap between modes B and C increases from 10.5 nm to 18.5 nm with increasing etching depth. It is an indication of enhanced feedback coupling and, as a result, lower threshold gain is expected and consistent with our observation of lower threshold pumping power. Figure 6b shows the below-threshold spectra for 350-nm-deep but varying period devices (E1 to E5). The bandgap separation is almost equal for five devices with fixed etching depth but different lattice period. The feedback coupling is nearly independent of lattice period.

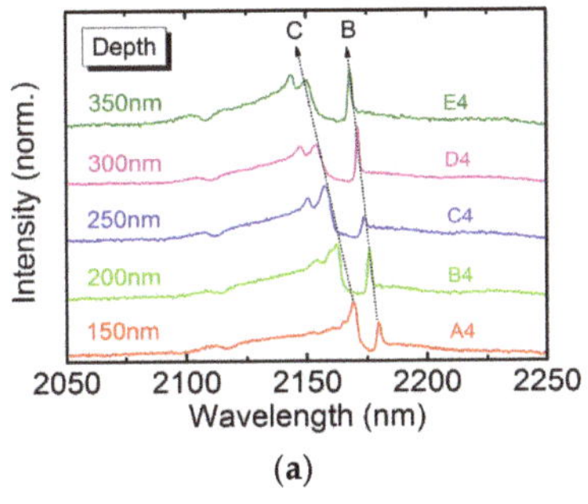
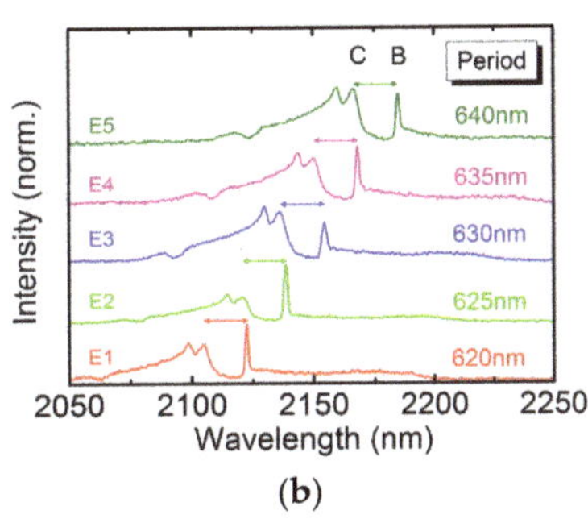

Figure 6. The below-threshold emission spectra for devices with (**a**) 635-nm-period but varying etching depth (A4 to E4) and (**b**) 350-nm-deep but varying period (E1 to E5).

The identified band-edge modes are normalized in frequency by dividing lattice periods by vacuum wavelengths and plot against etching depths as shown in Figure 7a. Only 630- and 635-nm-period devices are plotted for clarity. Figure 7b shows their normalized lasing and Bragg frequencies versus etching depths. The lasing peaks are originated from band-edge mode B, which is the same as revealed in our previous work [6,11]. Regarding the Bragg frequencies, these are determined as the center of band-edge modes B and C. Therefore, the normalized lasing frequency is detuned from the Bragg condition by a detuning parameter ($\delta < 0$), whose magnitude increases with increasing etching depth.

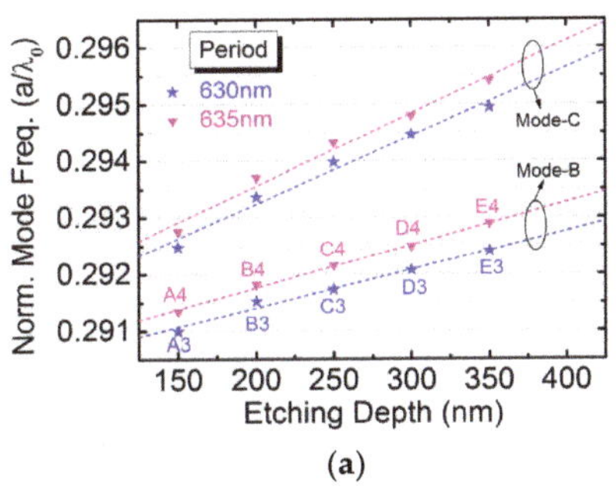
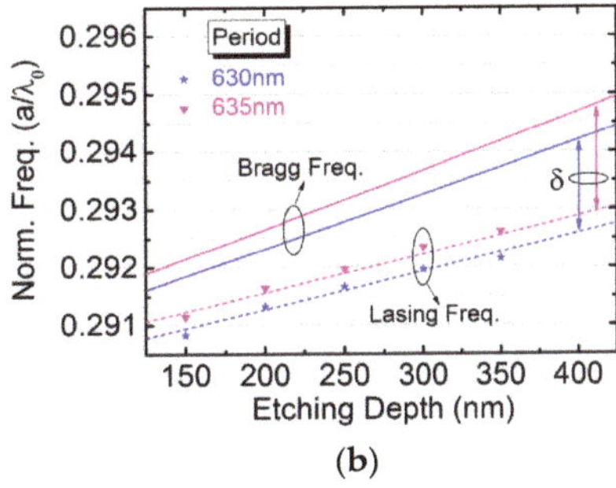

Figure 7. (**a**) The normalized band-edge mode frequency as well as (**b**) the normalized lasing and Bragg frequency is plotted against etching depth for 630- and 635-nm-period devices.

Assuming that the logarithmic relationship between threshold gain (gth) and current density holds for optical pumping, its inverse relationship,

$$P_{th}(g_{th}) = P_0 exp\left(\frac{g_{th} - g_0}{g_0}\right),$$

is modeled by two parameters of reference gain (g_0) and associated pumping density (P_0). Let the offsets in Figures 4b and 5a correspond to the same minimum achievable modal gain (say $g_0 = 10 \text{ cm}^{-1}$) for PCSEL devices. Therefore, the net modal gain can be extracted, while the detuning is extracted from Figure 6b. Figure 8 show the threshold modal gain as a function of normalized frequency detuning. The vertical axis is relative to the reference gain. As the etching depth increases, the feedback coupling is enhanced to result in larger detuning but a lower threshold gain. The gain-detuning relationship is experimentally determined to be exponential-like and similar to theoretical predictions for the first-order modes of one-dimensional (1D) distributed feedback (DFB) lasers [12].

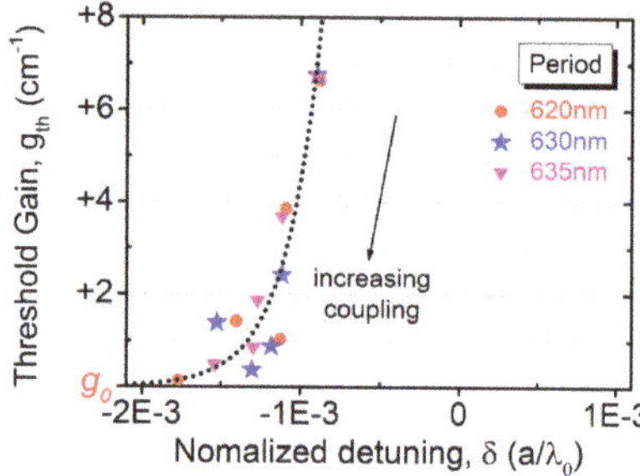

Figure 8. The threshold modal gain is plotted as a function of normalized frequency detuning.

4. Conclusions

In this paper, we have systematically studied the device characteristics of GaSb-based Mid-IR PCSELs with respect to etching depth and lattice period. In terms of lasing wavelength, the wavelength shift is about +31.4 nm for every 10-nm increase in period and −5 nm for 100-nm increase in depth. Measurement of below-threshold emission spectra identifies the bandgap as well as band-edge modes. The bandgap separation, which is a function of feedback coupling, increases with increasing depth but is independent of the lattice period. The criteria are set to select devices for threshold gain analysis. With increasing depth, the threshold pumping density decreases exponentially to a saturation level, which is assigned to the minimum device modal gain. The relative threshold gain is then plotted as a function of normalized frequency detuning. The gain-detuning relationship of PCSELs is similar to that of 1D DFB lasers.

Author Contributions: Z.-L.L. and S.-C.L. conceived and designed the experiments; S.-C.L. and H.-W.C. performed the experiments; Z.-L.L., S.-C.L., and G.L. analyzed the data; K.-W.S. and C.-P.L. contributed reagents/materials/analysis tools; Z.-L.L. and G.L. wrote the paper.

Acknowledgments: This research was funded by the Ministry of Science and Technology under grant number MOST 107-2218-E-009-034. The APC was funded by the Ministry of Science and Technology under grant number MOST 106-2221-E-009-121-MY3. The authors appreciate the services and facilities provided by the Center for Nano Science and Technology (CNST) of National Chiao Tung University. They also thank Dr. Chien-Hung Pan of Truelight Corporation for valuable discussions.

Conflicts of Interest: The authors declare no conflict of interest.

References

1. Schilt, S.; Vicet, A.; Werner, R.; Mattiello, M.; Thévenaz, L.; Salhi, A.; Rouillard, Y.; Koeth, J. Application of antimonide diode lasers in photoacoustic spectroscopy. *Spectrochim. Acta A* **2004**, *60*, 3431–3436. [CrossRef] [PubMed]
2. Yin, Z.; Tang, X. A review of energy bandgap engineering in III–V semiconductor alloys for mid-infrared laser applications. *Solid State Electron.* **2007**, *51*, 6–15. [CrossRef]
3. Bauer, A.; Rößner, K.; Lehnhardt, T.; Kamp, M.; Höfling, S.; Worschech, L.; Forchel, A. Mid-infrared semiconductor heterostructure lasers for gas sensing applications. *Semicon. Sci. Technol.* **2011**, *26*, 014032. [CrossRef]
4. Hirose, K.; Liang, Y.; Kurosaka, Y.; Watanabe, A.; Sugiyama, T.; Noda, S. Watt-class high-power, high-beam-quality photonic-crystal lasers. *Nat. Photon.* **2014**, *8*, 406–411. [CrossRef]
5. Pan, C.H.; Lin, C.H.; Chang, T.Y.; Lu, T.C.; Lee, C.P. GaSb-based mid infrared photonic crystal surface emitting lasers. *Opt. Express* **2015**, *23*, 11741–11747. [CrossRef] [PubMed]
6. Hsu, M.-Y.; Lin, G.; Pan, C.-H. Electrically injected 1.3 um quantum-dot photonic-crystal surface-emitting lasers. *Opt. Express* **2017**, *25*, 32697–32704. [CrossRef]
7. Li, Z.L.; Chang, B.H.; Lin, C.H.; Lee, C.P. Dual-wavelength GaSb-based mid infrared photonic crystal surface emitting lasers. *J. Appl. Phys.* **2018**, *123*, 093102. [CrossRef]
8. Cheng, H.-W.; Lin, S.-C.; Li, Z.-L.; Sun, K.-W.; Lee, C.-P. PCSEL Performance of Type-I InGaAsSb Double-QWs Laser Structure Prepared by MBE. *Materials* **2019**, *12*, 317. [CrossRef] [PubMed]
9. Chen, T.S.; Li, Z.L.; Hsu, M.Y.; Lin, G.; Lin, S.D. Photonic crystal surface emitting lasers with quantum dot active region. *J. Lightwave Technol.* **2017**, *35*, 4547–4552. [CrossRef]
10. Li, Z.-L.; Kan, Y.-C.; Lin, G.; Lee, C.-P. Mid-Infrared Photonic-Crystal Surface-Emitting Lasers with InGaAs/GaAsSb 'W'-Type Quantum Wells Grown on InP Substrate. *Photonics* **2019**, *5*, 32. [CrossRef]
11. Hsu, M.Y.; Lin, G.; Li, Z.L.; Pan, P.C. Threshold current temperature dependence of quantum-dot photonic crystal surface-emitting lasers with respect to gain-cavity detuning. *Opt. Express* **2018**, *26*, 13483–13488. [CrossRef] [PubMed]
12. Kogelnik, H.; Shank, C.V. Coupled-Wave Theory of Distributed Feedback Lasers. *J. Appl. Phys.* **1972**, *43*, 2327–2335. [CrossRef]

 *micromachines* 

Article

Design and Fabrication of a Wavelength-Selective Near-Infrared Metasurface Emitter for a Thermophotovoltaic System

Atsushi Sakurai * and Yuki Matsuno

Niigata University, Niigata 950-2181, Japan; t11m063b@gmail.com
* Correspondence: sakurai@eng.niigata-u.ac.jp; Tel.: +81-25-262-7004

Received: 30 November 2018; Accepted: 22 February 2019; Published: 25 February 2019

Abstract: In this study, a tungsten-SiO_2-based metal–insulator–metal-structured metasurface for the thermal emitter of the thermophotovoltaic system was designed and fabricated. The proposed emitter was fabricated by applying the photolithography method. The fabricated emitter has high emissivity in the visible to near-infrared region and shows excellent wavelength selectivity. This spectral emissivity tendency agreed well with the result calculated by the finite-difference time-domain method. Additionally, the underlying mechanism of its emission was scrutinized. Study of the fabrication process and theoretical mechanisms of the emission, clarified in this research, will be fundamental to design the wavelength-selective thermal emitter.

Keywords: near-infrared light; wavelength-selective emitter; metasurface; photolithography method

1. Introduction

Recently, with global warming and the exhaustion of fossil fuels, there is an urgency to adapt an efficient way of utilizing renewable energy. Among the numerous renewable energies, solar energy is one of the most promising candidates. As an efficient method to convert solar energy into electricity, the thermophotovoltaic (TPV) system is attracting increasing attention [1,2]. The TPV system can efficiently convert thermal energy from any kind of heat source into electricity by absorbing the incident thermal energy with a wavelength selective absorber and emitting that energy with a thermal emitter, which emits photons in a certain wavelength that matches the highest quantum efficiencies (QE) of a specific photovoltaic (PV) cell. Moreover, the TPV system is environmentally friendly and discharges hardly any harmful substances or noise.

In the TPV system, the wavelength-selective thermal emitter plays a significant role in its energy efficiency [3]. The ideal thermal emitter requires high emissivity at the wavelength that matches the highest QE of a PV cell, and low thermal emissivity in the infrared region to reduce radiative heat loss. Additionally, its emissivity should be independent from the polarization and incident angles. To obtain a high-efficiency emitter, a metal–insulator–metal (MIM)-structured metasurface is employed for this study. MIM metasurfaces have been successfully developed for tailoring thermal radiation [4–8]. Conventional MIM metasurfaces were designed for room temperature applications. Therefore, MIM metasurfaces are usually fabricated with noble metals. But recent studies successfully demonstrated MIM metasurfaces for high temperature applications [9,10]. Tungsten (W) is a promising material for high temperature applications because it has a high melting point. However, the previous studies of such MIM metasurfaces with W were very few because of the difficulty in patterning W nano patterning. Lin et al. [11] computationally designed a W-based anisotropic metamaterial for a broadband absorber, but the designed nano-structure was very complicated. Lefebvre et al. [12] reported the experimental realization of complementary metal-oxide-semiconductor (CMOS)-compatible MIM perfect absorbers, of which their target wavelength is in the mid-infrared range. Therefore, it is necessary to improve the fabrication technology and the optical characteristics of near-infrared metasurfaces for high temperature applications.

In this study, we computationally design and experimentally fabricate a near-infrared metasurface emitter for a GaSb PV cell. To unravel the underlying resonance mechanisms and evaluate the measurement results, electromagnetic (EM) simulations based on the finite-difference time-domain (FDTD) method and an LC-circuit model are performed and compared with the measurement results.

2. Computational Design and Experimental Process

Figure 1a shows a schematic of the proposed emitter (i.e., a periodic metallic disk pattern on a dielectric film, which is placed on top of a metallic film). θ and ϕ represent the incident and polarization angles of the incident light, respectively. SiO_2 was chosen as a dielectric spacer and W was chosen as the metallic part of the proposed metasurface. Although there is a mismatch in the thermal expansion coefficient between W and SiO_2, a metasurface using these materials, which is stable up to 800 K, has been fabricated previously and reported [9]. Therefore, we assumed that our proposed MIM-structured metasurface was reasonable, to be utilized, for the TPV emitter, which requires a high operating temperature. Periods of the unit cell for the x and y directions were λ = 600 nm and the diameter of the W disks was w = 350 nm. The height of the W disks and thickness of the dielectric spacer were fixed to 50 and 100 nm, respectively. To compute the proposed structure's optical characteristics, we employed the Lumerical FDTD software. Dielectric functions of SiO_2 and W were obtained from the tabulated data from Palik [13].

Figure 1b represents the spectral emissivity calculated from normal reflectance obtained from Lumerical FDTD software (blue dot) for the proposed metasurface and the QE of the GaSb PV cell (orange dot) [14]. Here, since the W substrate is opaque, spectral emissivity (ε_λ) could be calculated from the Kirchhoff's law (i.e., $\varepsilon_\lambda = 1 - R_\lambda$, where R_λ is the spectral reflectance). It can be seen that the higher emissivity spectrum matched the higher QE region.

Figure 2 shows the fabrication process of the proposed metasurface; it was composed primarily of six steps: (a) Thin W and SiO_2 films were sputtered on a Si plate; (b) coating the top SiO_2 film with a positive photoresist; (c) pattern transfer to the photoresist with the exposure method using an i-line stepper; (d) after the post exposure bake, the photoresist layer was developed to form a resist pattern; (e) W was sputtered on the sample; and (f) the rest of the photoresist was removed.

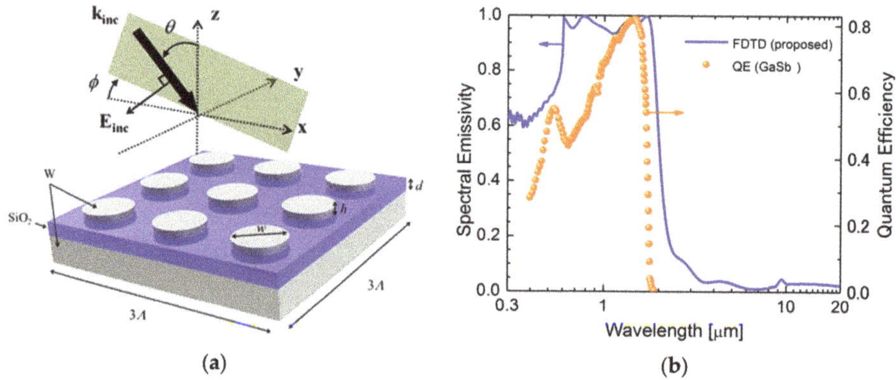

Figure 1. (a) Schematic of the proposed metasurface. (b) Simulated spectral emissivity of the proposed emitter obtained from finite-difference time-domain (FDTD) simulation (blue dot) and the quantum efficiencies (QE) of the GaSb PV cell (orange dot) [14].

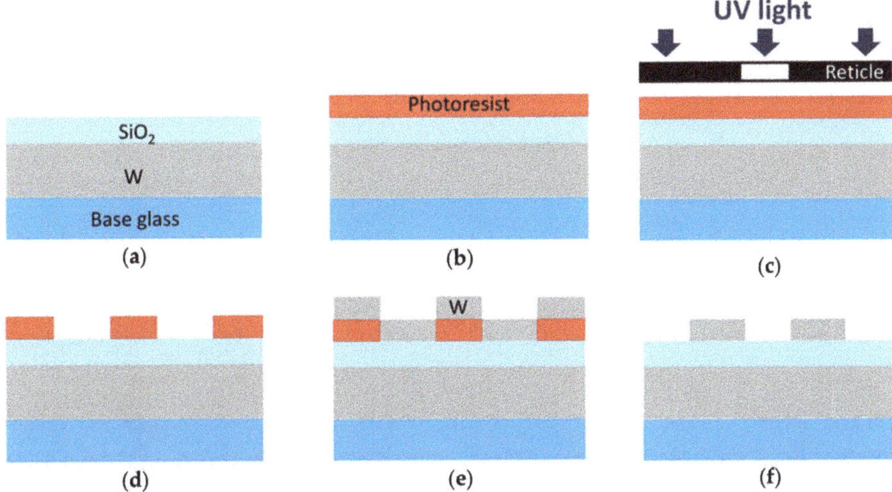

Figure 2. Schematic of the fabrication process of the proposed metasurface. (**a**) Thin W and SiO_2 films were sputtered; (**b**) photoresist coating; (**c**) ultraviolet (UV) light exposure through a reticle; (**d**) the photoresist layer was developed to form a resist pattern; (**e**) the thin W layer was sputtered; (**f**) lift-off (photoresist with the unnecessary W removed). Reproduced with permission from [8], published by OSA Publishing, 2017.

Both the base W film and SiO_2 film were sputtered for 100 nm on the substrate. Then, a positive type photoresist was spin-coated on the top SiO_2 film and prebaked before exposure to ultraviolet (UV) light to form the resist pattern. Reticle with a 0.375 µm hole was used for the i-line stepper. Afterwards, the exposed photoresist was dissolved and resist patterns appearred. Finally, W was sputtered for 50 nm on the sample and the unnecessary W of the rest of the photoresist was removed by acetone.

A fourier transform infrared spectroscopy (FTIR) spectrometer (Thermo Fisther Scientific iS50 FTIR Nicolet, Thermo Fisther Scientific, Waltham, MA, USA) and a visible spectrometer (Portable Spectral Solar Absorptance Measurement System PM-A2, Kouei Inc., Tokyo, Japan) were used to measure reflectivity for the infrared region and visible region, respectively. Once the reflectivity was obtained, the spectral directional emissivity could be obtained by applying Kirchhoff's law.

3. Results and Discussion

To evaluate the fabricated metasurface, its spectral emissivity was measured and compared to the calculation results of the FDTD method. Figure 3a represents the simulated spectral emissivity and the measured spectral directional emissivity of the fabricated metasurface (red dots). A visible spectrometer was utilized for the emissivity measurement up to 2.5 µm and FTIR was applied for the wavelength longer than 2.5 µm. The measured spectral emissivity showed reasonable agreement with the simulated spectral emissivity. However, there was an emissivity mismatch at the wavelength shorter than the cut-off wavelength. Figure 3b displays the top scanning electron microscopy (SEM) images of the fabricated metasurface. It can be seen that the W disks showed excellent symmetry. However, the diameters of the W disks were patterned slightly larger than as designed. This was the reason why the cut-off wavelength was red-shifted. In our future work, the pattern transfer technique will be improved.

Since the proposed emitter consists of a MIM structure, magnetic polariton (MP) could be excited in the dielectric spacer between the metals [15–17]. The emissivity peak at 1.7 µm, located near the highest QE region, was caused by the excitation of MP. Figure 4a shows the electromagnetic (EM) field distribution at the emissivity peak along the x–z plane at $y = 0$ nm. The color contour shows the logarithm of the normalized magnitude of the square of the y-component magnetic field. The vectors show the direction and magnitude of the electric field. At the peak wavelength, highly localized *y*-component magnetic field enhancement can be observed in the SiO$_2$ film between the W disk and the bottom W plate. Furthermore, the electric field created a closed current loop, which created an enhanced magnetic field and thus formed MP. Therefore, the proposed emitter—which excites MP—could be said to have a strong emissivity peak at the highest QE wavelength region of the PV cell.

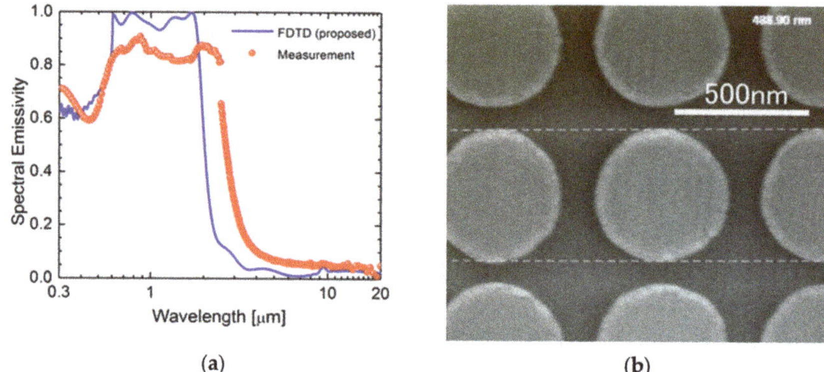

Figure 3. (**a**) Spectral emissivity of the proposed emitter obtained from FDTD simulation (blue line) and the fabricated sample (red dot). (**b**) Scanning electron microscopy (SEM) images of the fabricated metasurface from the top.

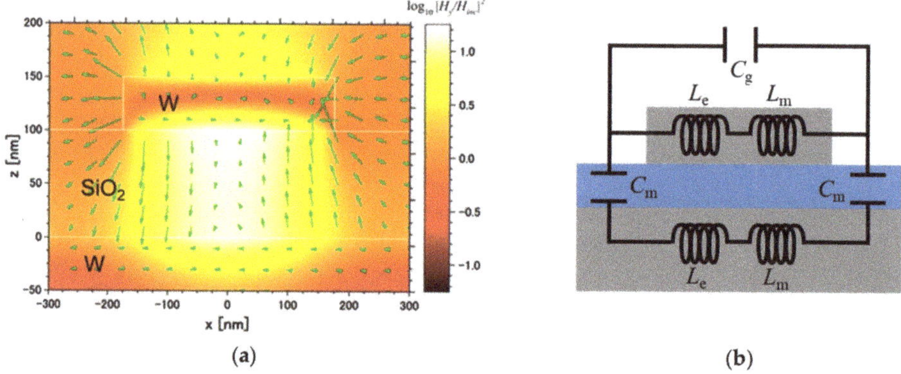

Figure 4. (**a**) Electromagnetic (EM) field profiles of the proposed emitter at 1.7 μm, calculated by FDTD simulation. The color counter shows the logarithm of the normalized magnitude of the square of the *y*-component magnetic field and the vectors show the direction and magnitude of the electric field; (**b**) equivalent LC circuit model.

As the theoretical method to predict the MP resonant wavelength, a LC circuit model can be applied to the proposed metasurface. Figure 4b describes the LC circuit model applied for the proposed emitter structure. Here, inductance and capacitance are represented as follows [15–17]:

$$L_{e,W} = -\frac{w}{\varepsilon_0 \omega^2 w \delta_W} \frac{\varepsilon'_W}{\left(\varepsilon'^2_W + \varepsilon''^2_W\right)} \tag{1}$$

$$L_{m,W} = \frac{1}{2}\mu_0 d \tag{2}$$

$$C_{g,W} = \frac{\varepsilon_0 h w}{\Lambda - w} \tag{3}$$

$$C_{m,SiO_2} = c_{1,W} \varepsilon'_{SiO_2} \varepsilon_0 \frac{w^2}{d} \tag{4}$$

$L_{e,W}$ and $L_{m,W}$ are the kinetic and magnetic inductance of W. $C_{g,W}$ is used to approximate the gap capacitance between the W disks. C_{m,SiO_2} means the parallel-plate capacitance. μ_0 is the permeability of the vacuum and ω is the angular frequency. ε'_W and ε''_W represent the real and imaginary parts of the dielectric function of W. ε_0 and ε'_{SiO_2} are the dielectric function of the vacuum and SiO_2, respectively, and $c_{1,W} = 0.32$ is the numerical factor to consider the fringe effect or non-uniform charge distribution along the surface of the capacitor. Originally, the numerical factor is recommended to be used in the range between about 0.2 and 0.3 [15]. $\delta_W = \frac{\lambda}{2\pi\kappa_w}$ is the effective penetration depth of W, where κ_W is the extinction coefficient of W. The total impedance of this LC circuit model can be obtained as:

$$Z_{tot}(\omega) = \frac{L_{m,W} + L_{e,W}}{1 - \omega^2 C_{g,W}(L_{m,W} + L_{e,W})} - \frac{2}{\omega^2 C_{m,SiO_2}} + L_{m,W} + L_{e,W} \tag{5}$$

Here, note that the dielectric function and penetration depth were wavelength-dependent, and the impedance was a module (i.e., no possible dephasing effects were considered). The resonance conditions of MP could be obtained by zeroing the total impedance (i.e., $Z_{tot} = 0$). For the proposed emitter, its resonant wavelength was predicted as 1.7 μm. This predicted wavelength was reasonably close to the peak wavelength of the designed emitter.

The spectral emissivity at the oblique incident waves, up to 30° for Transverse Magnetic (TM) and Transverse Electric (TE) waves, was calculated. The spectral emissivity can be obtained from Kirchhoff's law (i.e., $\varepsilon_\lambda = 1 - R_\lambda$), where the reflectance R was calculated by the FDTD. To clarify the angle-dependent behavior, the wideband analysis at oblique incidence was performed by using the FDTD algorithm suggested by Liang et al. [18].

The insensibility of the spectral emissivity to the incident and polarization angles was essential for the efficient wavelength-selective emitter. Figure 5 is the contour plot of the spectral emissivity of the proposed emitter obtained from FDTD simulation: (a) at TM waves; and (b) at TE waves, in terms of wavelength and incident angle. It can be observed that for both TM and TE waves, the proposed emitter maintained high emissivity at the high QE wavelength region, while emissivity in the infrared region was low in the wide incident angles. The emissivity peak around 1.7 μm originated from MP. On the other hand, the emissivity peak around 0.7 μm was because of the excitation of surface plasmon [17]. It can be seen that surface plasmon had a directional dependence, but the MP resonance frequency was insensitive to the polar angle.

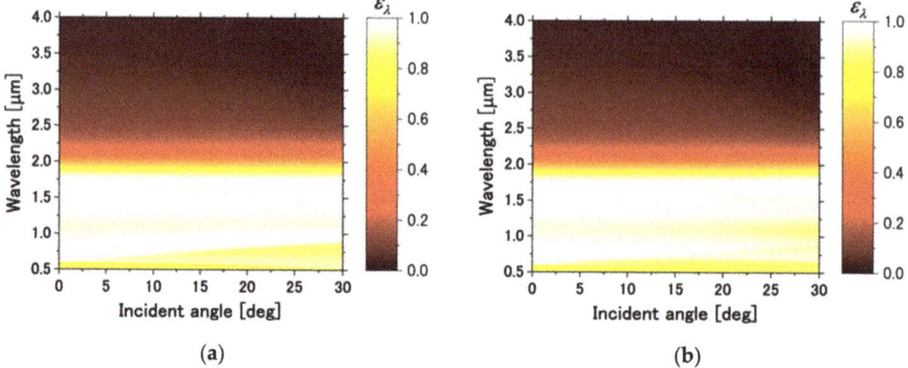

Figure 5. Contour diagram of the emissivity of the proposed emitter obtained by FDTD simulation for (**a**) Transverse Magnetic (TM) waves and (**b**) Transverse Electric (TE) waves, in terms of wavelength and incident angle up to 30°.

Finally, ideal efficiency of the TPV system, as a function of PV cell temperature, was predicted based on theoretical calculation [19] and available experimental data. Figure 6a shows the ideal emissivity spectra with wavelength ranges of 0.25 μm (case 1), 0.50 μm (case 2), and 0.75 μm (case 3). Here, the peak position was located at the highest quantum efficiency region. For comparison, the present emissivity spectra, ε_λ obtained by the numerical simulation, the experimental measurement, and blackbody were used to predict the efficiency of the TPV system.

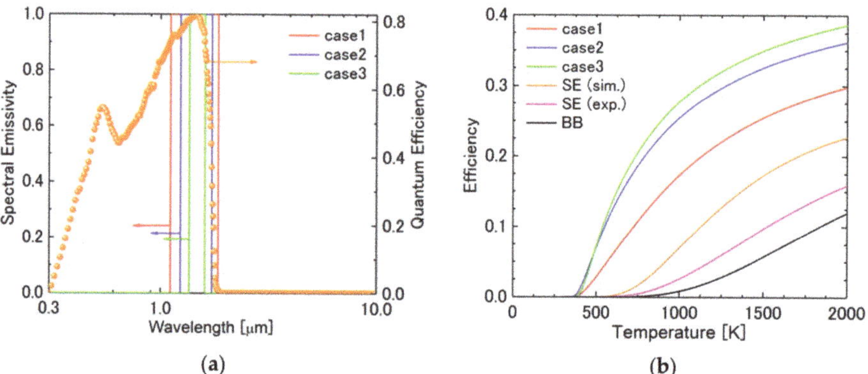

Figure 6. (**a**) Ideal emissivity spectra with wavelength ranges of 0.25 μm (case 1), 0.50 μm (case 2), and 0.75 μm (case 3). The peak position was located at the highest quantum efficiency region. (**b**) Efficiency of the present TPV system with emissivity spectra based on the three cases, the measured and simulated emissivity, and blackbody.

The input power, P_in, was defined by using Planck's spectral distribution of emissive power.

$$P_\text{in} = \int \frac{2\pi\varepsilon_\lambda hc^2}{\lambda^5\left(exp\left[\frac{hc}{\lambda kT}\right] - 1\right)} d\lambda \tag{6}$$

where h is the Planck constant, c is the speed of light, λ is wavelength, and k is the Boltzmann constant. The maximum power P_m is defined in a manner similar to a classic derivation by Loferski [20].

$$P_m = \frac{q J_{sc} (V_{mp})^2}{(kT + q V_{mp})} \tag{7}$$

where q is the electron charge, J_{sc} is the short-circuit current, and V_{mp} is the voltage at maximum power. In the ideal case, we assumed that there was no heat loss for the input power, and the view factor from the emitter to the PV cell was equal for unity. Therefore, the efficiency of the TPV system was defined as:

$$\eta = \frac{P_m}{P_{in}} \tag{8}$$

Figure 6b show the efficiency of the present TPV system with emissivity spectra based on the three cases, the measured and simulated emissivity, and blackbody. The wavelength-selective emitter (SE) had better results than the blackbody (BB) emitter, because the blackbody emitter had a large, undesired thermal radiation in the mid-infrared range. By the simulated emissivity spectra of the SE, when the emitter temperature exceeded about 1600 K, the efficiency approached 20%. However, the efficiency with the experimentally demonstrated SE was about half of the simulated one, because there was a mismatch between the emissivity spectra with the experiment and the quantum efficiency. Next, the three ideal emissivity spectra were introduced to show the theoretical maximum efficiency of this system. In case 1, the wavelength range was the narrowest, hence, the maximum efficiency was expected. From the calculated results, the efficiency of case 1 approached 40% when the emitter temperature became about 2000 K. In case 2 and 3, with broader wavelength ranges, the efficiency decreased, as compared with case 1. It was also true that there was emissivity mismatch with high quantum efficiency area.

4. Conclusion

The MIM-structured W-SiO$_2$-based metasurface emitter for the TPV system, which shows a high emissivity peak at the high QE wavelength region of the PV cell, was designed and fabricated. The designed metasurface was fabricated by applying the photolithography method and it was compared to FDTD simulation results. The fabricated metasurface showed high emissivity in the visible to near-infrared light region and indicated excellent wavelength-selectivity. Its emissivity tendency agreed reasonably well with the FDTD simulation. In addition, we clarified that the emissivity enhancement that is underpinning the emissivity peak at 1.7 μm originated from the excitation of MP by calculating the EM field and LC circuit model. It has been shown that the spectral absorption of the proposed emitter is nearly independent of the incident and polarization angles. This study will not only help to understand the mechanisms that can be used to tailor the emissivity enhancement, but will also facilitate the design and practical fabrication of nanostructures for applications in TPV systems.

Author Contributions: Conceptualization, A.S.; Funding acquisition, A.S.; Investigation, A.S. and Y.M.; Project administration, A.S.; Supervision, A.S.

Acknowledgments: This work was funded by JSPS KAKENHI (grant numbers 15K17985, 18K03974). The authors would like to thank Masahide Akiyama, Yoshihide Ikegami, and Kunihiko Murakami at KYOCERA Corporation for sample fabrication and Tsuyoshi Totani at Hokkaido University for reflectance measurement.

Conflicts of Interest: The authors declare no conflicts of interest.

References

1.   Bermel, P.; Ghebrebrhan, M.; Chan, W.; Yeng, Y.X.; Araghchini, M.; Hamam, R.; Marton, C.H.; Jensen, K.F.; Soljacic, M.; Joannopoulos, J.D.; et al. Design and global optimization of high-efficiency thermophotovoltaic systems. *Opt. Express* **2010**, *18*, A314–A334. [CrossRef] [PubMed]

2. Lenert, A.; Bierman, D.M.; Nam, Y.; Chan, W.R.; Celanovic, I.; Soljacic, M.; Wang, E.N. A nanophotonic solar thermophotovoltaic device. *Nat. Nanotechnol.* **2014**, *9*, 126–130. [CrossRef] [PubMed]

3. Nam, Y.; Yeng, Y.X.; Lenert, A.; Bermel, P.; Celanovic, I.; Soljacic, M.; Wang, E.N. Solar thermophotovoltaic energy conversion systems with two-dimensional tantalum photonic crystal absorbers and emitters. *Sol. Energy Mater. Sol. Cells* **2014**, *122*, 287–296. [CrossRef]

4. Landy, N.I.; Sajuyigbe, S.; Mock, J.J.; Smith, D.R.; Padilla, W.J. Perfect Metamaterial Absorber. *Phys. Rev. Lett.* **2008**, *100*, 207402. [CrossRef] [PubMed]

5. Aydin, K.; Ferry, V.E.; Briggs, R.M.; Atwater, H.A. Broadband polarization-independent resonant light absorption using ultrathin plasmonic super absorbers. *Nat. Commun.* **2011**, *2*, 517. [CrossRef] [PubMed]

6. Sakurai, A.; Zhao, B.; Zhang, Z.M. Resonant frequency and bandwidth of metamaterial emitters and absorbers predicted by an RLC circuit model. *J. Quant. Spectrosc. Radiat. Transfer* **2014**, *149*, 33–40. [CrossRef]

7. Sakurai, A.; Zhao, B.; Zhang, Z.M. Effect of polarization on dual-band infrared metamaterial emitters or absorbers. *J. Quant. Spectrosc. Radiat. Transfer* **2015**, *158*, 111–118. [CrossRef]

8. Matsuno, Y.; Sakurai, A. Perfect infrared absorber and emitter based on a large-area metasurface. *Opt. Mater. Express* **2017**, *7*, 618–626. [CrossRef]

9. Han, S.; Shin, J.-H.; Jung, P.-H.; Lee, H.; Lee, B.J. Broadband Solar Thermal Absorber Based on Optical Metamaterials for High-Temperature Applications. *Adv. Opt. Mater.* **2016**, *4*, 1265–1273. [CrossRef]

10. Wang, H.; Alshehri, H.; Su, H.; Wang, L. Design, fabrication and optical characterizations of large-area lithography-free ultrathin multilayer selective solar coatings with excellent thermal stability in air. *Sol. Energy Mater. Sol. Cells* **2018**, *174*, 445–452. [CrossRef]

11. Lin, Y.; Cui, Y.; Ding, F.; Fung, K.H.; Ji, T.; Li, D.; Hao, Y. Tungsten based anisotropic metamaterial as an ultra-broadband absorber. *Opt. Mater. Express* **2017**, *7*, 606–617. [CrossRef]

12. Lefebvre, A.; Costantini, D.; Doyen, I.; Levesque, Q.; Lorent, E.; Jacolin, D.; Greffet, J.J.; Boutami, S.; Benisty, H. CMOS compatible metal-insulator-metal plasmonic perfect absorbers. *Opt. Mater. Express* **2016**, *6*, 2389–2396. [CrossRef]

13. Palik, E.D. *Handbook of Optical Constants of Solids*; Vol. 3, Academic Press: San Diego, CA, USA, 1998.

14. JX Crystals, Inc. Available online: http://jxcrystals.com/drupal/GaSb (accessed on 30 September 2018).

15. Lee, B.J.; Wang, L.P.; Zhang, Z.M. Coherent thermal emission by excitation of magnetic polaritons between periodic strips and a metallic film. *Opt. Express* **2008**, *16*, 11328–11336. [CrossRef] [PubMed]

16. Wang, L.P.; Zhang, Z.M. Wavelength-selective and diffuse emitter enhanced by magnetic polaritons for thermophotovoltaics. *Appl. Phys. Lett.* **2012**, *100*, 063902. [CrossRef]

17. Zhao, B.; Wang, L.P.; Shuai, Y.; Zhang, Z.M. Thermophotovoltaic emitters based on a two-dimensional grating/thin-film nanostructure. *Int. J. Heat Mass Transfer* **2013**, *67*, 637–645. [CrossRef]

18. Liang, B.; Bai, M.; Ma, H.; Ou, N.; Miao, J. Wideband analysis of periodic structures at oblique incidence by material independent FDTD algorithm. *IEEE Trans. Antennas Propag.* **2013**, *62*, 354–360. [CrossRef]

19. Ferguson, L.G.; Fraas, L.M. Theoretical study of GaSb PV cells efficiency as a function of temperature. *Sol. Energy Mater. Sol. Cells* **1995**, *39*, 11–18. [CrossRef]

20. Loferski, J.J. Theoretical considerations governing the choice of the optimum semiconductor for photovoltaic solar energy conversion. *J. Appl. Phys.* **1956**, *27*, 777–784. [CrossRef]

 *micromachines*

Article

Development of a New Laparoscopic Detection System for Gastric Cancer Using Near-Infrared Light-Emitting Clips with Glass Phosphor

Shunko A. Inada [1,*]**, Hayao Nakanishi** [2]**, Masahiro Oda** [3]**, Kensaku Mori** [3,4]**, Akihiro Ito** [5]**, Junichi Hasegawa** [6]**, Kazunari Misawa** [7] **and Shingo Fuchi** [8]

[1] Department of Mechanical Science and Engineering, Faculty of Science and Technology, Hirosaki University, 3 Bunkyo-cho, Hirosaki 036-8561, Japan

[2] Laboratory of Pathology and Clinical Research, Aichi Cancer Center Aichi Hospital, Okazaki 444-0011, Japan; hnakanis@aichi-cc.jp

[3] Graduate School of Informatics, Nagoya University, Nagoya 464-8601, Japan; moda@mori.m.is.nagoya-u.ac.jp (M.O.); kensaku@is.nagoya-u.ac.jp (K.M.)

[4] Research Center for Medical Bigdata, National Institute of Informatics, Tokyo 101-8430, Japan

[5] Graduate School of Pharmaceutical Sciences, Meijo University, Nagoya 468-0073, Japan; itou.pha@chubuh.johas.go.jp

[6] Department of Media Engineering, School of Engineering, Chukyo University, Nagoya 466-8666, Japan; hasegawa@sist.chukyo-u.ac.jp

[7] Department of Gastroenterological Surgery, Aichi Cancer Center Hospital, Nagoya 464-8681, Japan; misawakzn@aichi-cc.jp

[8] Department of Electrical Engineering and Electronics, College of Science and Engineering, Aoyama Gakuin University, Kanagawa 252-5258, Japan; fuchi@ee.aoyama.ac.jp

* Correspondence: inada@hirosaki-u.ac.jp; Tel.: +81-172-36-2111

Received: 19 December 2018; Accepted: 19 January 2019; Published: 24 January 2019

Abstract: Laparoscopic surgery is now a standard treatment for gastric cancer. Currently, the location of the gastric cancer is identified during laparoscopic surgery via the preoperative endoscopic injection of charcoal ink around the primary tumor; however, the wide spread of injected charcoal ink can make it difficult to accurately visualize the specific site of the tumor. To precisely identify the locations of gastric tumors, we developed a fluorescent detection system comprising clips with glass phosphor (Yb^{3+}, Nd^{3+} doped to Bi_2O_3-B_2O_3-based glasses, size: 2 mm × 1 mm × 3 mm) fixed in the stomach and a laparoscopic fluorescent detection system for clip-derived near-infrared (NIR) light (976 nm). We conducted two ex vivo experiments to evaluate the performance of this fluorescent detection system in an extirpated pig stomach and a freshly resected human stomach and were able to successfully detect NIR fluorescence emitted from the clip in the stomach through the stomach wall by the irradiation of excitation light (λ: 808 nm). These results suggest that the proposed combined NIR light-emitting clip and laparoscopic fluorescent detection system could be very useful in clinical practice for accurately identifying the location of a primary gastric tumor during laparoscopic surgery.

Keywords: gastric cancer; laparoscopic surgery; fluorescent clip; near-infrared fluorescence imaging

1. Introduction

Gastric cancer is a major cause of disease-related death worldwide, causing the deaths of 723,000 people in 2012 [1], and it is the most common type of cancer in East Asia [2]. Among several options for the treatment of gastric cancer, such as surgery, radiotherapy and chemotherapy, surgical resection is the most reliable. Although open surgery has long been the main form of treatment for gastric cancer, laparoscopic surgery is now recognized as an alternative surgical modality for gastric cancers. Compared with open surgery, laparoscopic surgery is less invasive, results in good quality of life for the patient, and is almost equally effective [3–5].

In conventional laparoscopic surgery for gastric cancer, charcoal ink (activated carbon) or India ink is endoscopically injected around the primary gastric tumor to enable the visualization of the location of the tumor in the stomach intraperitoneally [6,7]. However, it is sometimes difficult to visualize the exact location of the primary tumor during the laparoscopic operation, as the injected charcoal ink diffusely spreads to areas distant from the tumor in the stomach, unlike in tumors of the colon. This charcoal ink diffusion may result in resection with a broad margin containing wide tumor-free tissue, which results in great inconvenience for the patient. To overcome this problem, we developed a new convenient method with which to visualize the exact location of a primary gastric tumor from outside the stomach.

In the present study, we focused on the near-infrared (NIR) wavelength around the 1000 nm band, which has good biological transmission activity [8], and developed a new fluorescent chip with glass phosphor and a laparoscopic fluorescent detection system to assist in the performance of safe and accurate laparoscopic removal of gastric cancer. Two ex vivo experiments performed in this pilot study demonstrated the successful use of this new detection system.

2. Materials and Methods

2.1. Development of Glass Phosphor

To obtain a biological transmission light with a wavelength of around 1000 nm, we developed a glass phosphor in which Yb^{3+} and Nd^{3+} were doped to Bi_2O_3-B_2O_3-based glasses. The glass phosphor was synthesized by the melt-quenching method. Powders of Yb_2O_3, Nd_2O_3, Bi_2O_3, and B_2O_3 were mixed. The mixed powders were melted at 1250 °C in an Al_2O_3 crucible in an electric furnace. After 10 min, the molten liquid was poured into stainless steel-molded plates kept at room temperature to enable the formation of the glass. The glass phosphor could be excited by light with a wavelength of 808 nm. Figure 1 shows the photoluminescence spectrum of the glass phosphor.

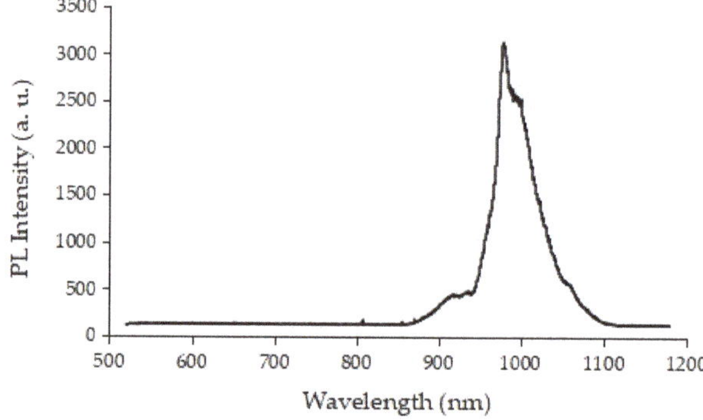

Figure 1. Photoluminescence spectrum of Yb^{3+} and Nd^{3+} co-doped Bi_2O_3-B_2O_3 glass phosphor. Light with a wavelength of 808 nm is suitable for excitation.

As light output is influenced by the structure of glass phosphor, we used the glass polishing method to create three shapes: rectangular (size: 1.8 mm × 1 mm × 3.5 mm), square pyramid array in two faces (size: 1.8 mm × 1 mm × 3.5 mm mm, pyramid size: 0.5 mm × 0.5 mm × 0.3 mm), and square pyramid array in four faces (size: 1.8 mm × 1 mm × 3.5 mm, pyramid size: 0.23 mm × 0.23 mm × 0.2 mm); we then compared the light outputs of the three shapes. The phosphor glass was excited using a semiconductor laser diode (SDL-808-LM-3000MFL; Shanghai Dream Lasers Technology Co., Ltd., Shanghai, China) with a wavelength of 808 nm, and the fluorescence intensity was measured using a power meter (PM100A power meter, S310C thermal power head, Thorlabs, Inc.,

Newton, NJ, USA) through a notch filter. The glass phosphor was irradiated with 1000 mW from a distance of 5 cm. Figure 2 shows the fluorescence intensity output of each glass phosphor structure. The square-pyramid-array-in-two-faces structure had the greatest intensity output (2.1 times greater than that of the rectangular structure). This is probably because the light emitted in the glass phosphor was refracted and focused on the large bottom area of the square pyramid.

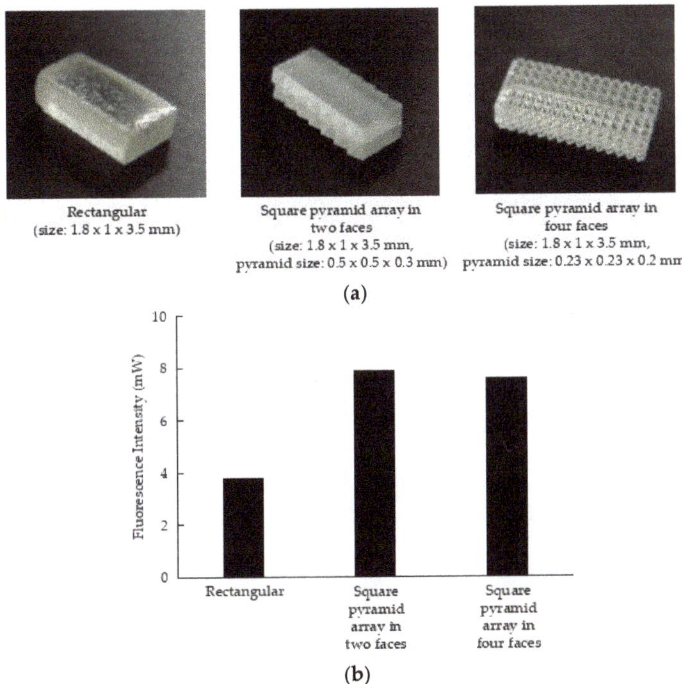

Figure 2. Morphological structures of the three glass phosphor shapes created and their respective fluorescence intensity outputs. (**a**): Rectangular, square-pyramid-array-in-two-faces and square-pyramid-array-in-four-faces structures were made using the glass polishing method. (**b**): The fluorescence intensity of each glass phosphor structure was measured at a distance of 5 cm and an irradiation intensity of 1000 mW using a semiconductor laser diode (808 nm). The square pyramid array with two faces showed the greatest intensity output.

2.2. Development of the Fluorescent Clip with Glass Phosphor

The fluorescent clip developed in this study consists of a commercially available hemostatic clip (HX-610-090L, Olympus Medical Systems Corp., Tokyo, Japan) as the main support and the glass phosphor with the square pyramid array in two faces structure at the tip. The glass phosphor was attached to the tip of the hemostatic clip with an adhesion bond (Figure 3).

2.3. Development of the Laparoscopic Fluorescent Detection System

To detect the NIR light emitted from the fluorescent chip, we developed a laparoscopic fluorescent detection system composed of a medical rigid scope (WA53000A, Olympus Medical Systems Corp., Tokyo, Japan), two notch filters, and a NIR CCD camera with a CCTV lens and camera controller (camera: C10639-80; camera controller: C2741-62; Hamamatsu Photonics K.K., Hamamatsu, Japan). The camera unit is connected to a monitor and personal computer, which makes it possible to obtain the images in real-time and record pictures and videos. A semiconductor laser diode (SDL-808-LM-3000MFL; Shanghai Dream Lasers Technology Co., Ltd., Shanghai, China) with a wavelength of 808 nm was used to excite the fluorescent clip. Figure 4 shows an overview of the laparoscopic fluorescent detection system.

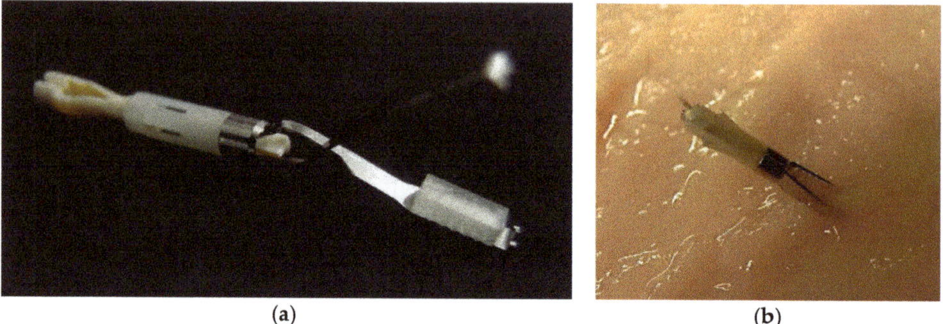

(a)   (b)

Figure 3. Photographs of (**a**) the fluorescent clip with glass phosphor at the tip and (**b**) a glass phosphor-attached clip fixed to the stomach mucosa.

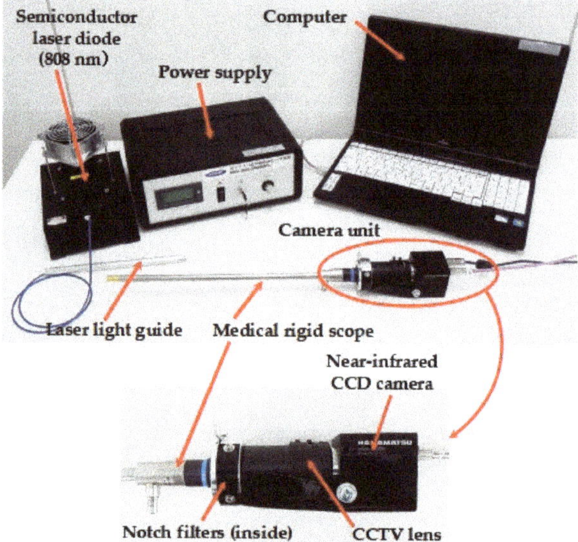

Figure 4. The laparoscopic fluorescent detection system composed of a medical rigid scope with a near-infrared CCD camera, a semiconductor laser diode (808 nm) as an excitation light source, and a computer with monitor.

2.4. Experimental Procedures

2.4.1. Ex Vivo Detection of the Fluorescent Clips in Pig Stomach

To evaluate the performance of the fluorescent clip and the laparoscopic fluorescent detection system, we carried out an ex vivo experiment using a commercially available pig stomach. To compare the efficacy of the glass phosphor structure, a fluorescent clip with the rectangular glass phosphor structure and one with the square-pyramid-array-in-two-faces structure were used. Each fluorescent clip was fixed on the gastric mucosa and covered with stomach wall. The rigid scope (camera unit) and the excitation light source were fixed at a distance of 5 cm from the outside of the stomach.

The excitation light was irradiated at 500, 1000, and 1500 mW of output power. The experiment was done using pig stomach sections with wall thicknesses of 3 and 10 mm. We carried out this kind of experiment using pig stomach in triplicate, and similar results were obtained.

2.4.2. Ex Vivo Detection of the Fluorescent Clips in Human Stomach

To further evaluate the performance of the fluorescent clip and the laparoscopic fluorescent detection system, we used a freshly resected stomach (thickness: 10 mm) from a patient with gastric cancer who underwent a total gastrectomy. The study protocol was approved by the institutional ethical review board of Aichi Cancer Center (approval number: H25-3-29; approval date: 2013.6.14) and written informed consent was obtained from the patient prior to the sample collection. A fluorescent clip (square-pyramid-array-in-two-faces, glass phosphor structure) was fixed on the gastric mucosa and covered with stomach wall. The rigid scope (camera unit) and the excitation light source were fixed at a distance of 6 cm from the stomach surface. The excitation light was irradiated at an output level of 1500 mW. The detection of fluorescence was done from the proximal and distal side of the stomach wall where the clip was fixed.

2.4.3. Evaluation of the Cytotoxicity of Glass Phosphor

The toxic side effect of the glass phosphor on cells was examined using normal human mesothelial cells. In this study, to examine the toxicity of this glass phosphor to the normal human tissue, we used immortalized, normal human mesothelial cells (HOMC) with stable growth potential established in our laboratory rather than cancer cells [9]. The cells (1×10^5 cells/dish) were seeded on a culture dish ($\varphi 3.5$ cm) and cultured in Dulbecco's modification of Eagle medium supplemented with 10% fetal bovine serum and antibiotics at 37 °C in a 5% CO_2 atmosphere. After 24 h, glass phosphor weights of 0, 0.5, 1.0, and 1.5 g were put into each dish and cultured for another 72 h. The cell numbers were counted with a hemocytometer in triplicate.

3. Results

3.1. Ex Vivo Detection of the Fluorescent Clips in Pig Stomach

Pig stomach was first used to evaluate the performance of the fluorescent clip and the laparoscopic fluorescent detection system. Figure 5 shows that the fluorescence intensity of the fluorescent clip increased depending on the output power of the excitation light. The glass phosphor with the square-pyramid-array-in-two-faces structure obtained a greater fluorescence output than the rectangular structure in each experimental condition. The structural effect was demonstrable when the clip was covered with a section of stomach with thicknesses of 3 and 10 mm. Thus, we decided to use the square-pyramid-array-in-two-faces, glass phosphor structure in the following investigations.

3.2. Ex Vivo Detection of the Fluorescent Clips in Human Stomach

Freshly resected human stomach with a thickness of 10 mm, which contained a significant amount of residual blood, was used to further evaluate the fluorescent clip and the laparoscopic fluorescent detection system in a more similar condition to the clinical setting. Figure 6b shows the detection of the fluorescent clip fixed in the stomach from the proximal and distal side of the stomach wall. The control images were obtained using a fluorescent clip fixed in the gastric mucosa without covering by the stomach. The fluorescence intensity detected from the proximal side of the stomach wall was greater than the fluorescence detected from the distal side of the stomach wall. However, sufficient fluorescence was still observed even from the distal side of the stomach wall Figure 6a, suggesting the possibility that the fluorescent clip would be effective in a clinical setting.

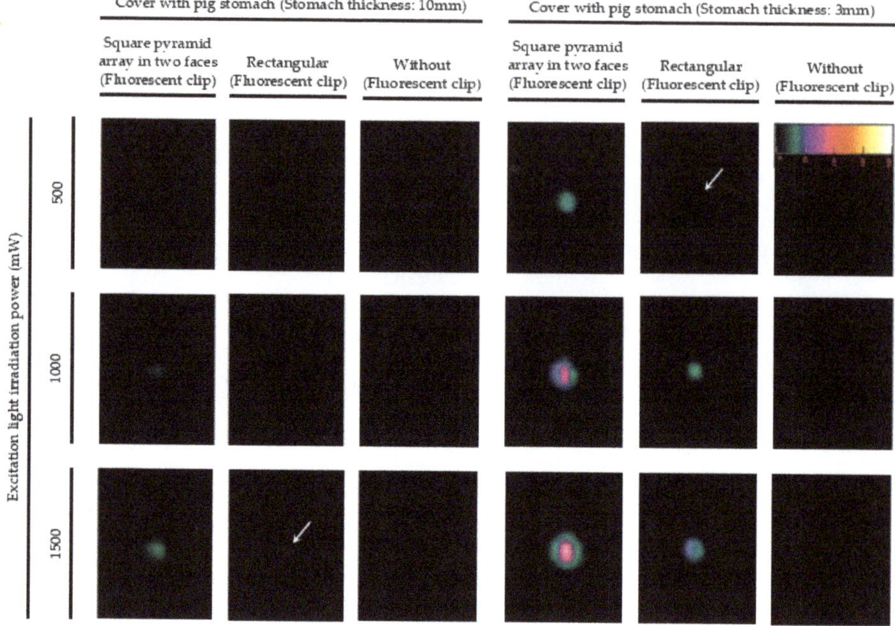

Figure 5. Detection of the fluorescent clips fixed in a commercially available pig stomach sample. The fluorescent image of the clip with the square-pyramid-array-in-two-faces, glass phosphor structure was successfully obtained with good fluorescence output.

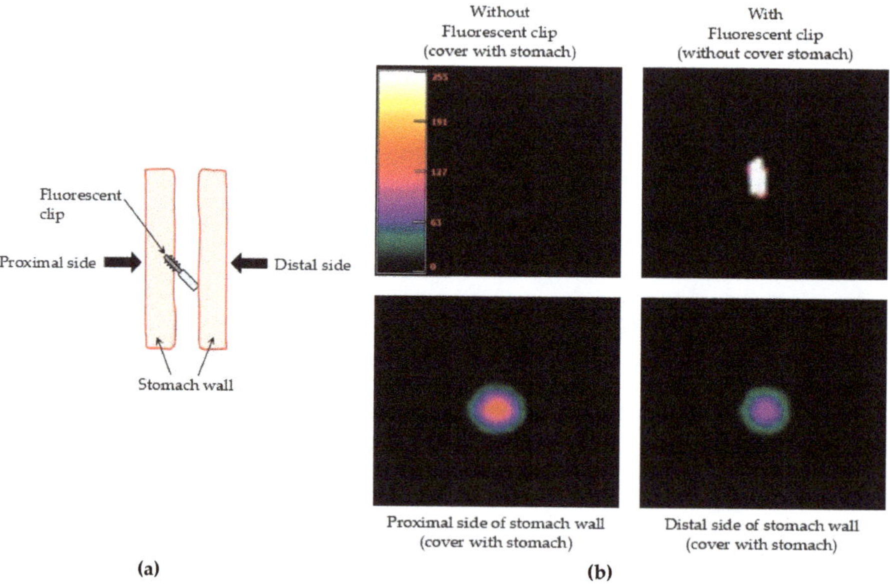

Figure 6. Ex vivo detection of the near-infrared fluorescent clips fixed in human stomach. (**a**) Schematic representation of the detection method of the fluorescent clips fixed in the human stomach. (**b**) A stronger fluorescent image of the clip from the proximal side of the stomach wall than from the distal side was observed.

3.3. Cytotoxicity of the Glass Phosphor

We examined the cytotoxicity of the glass phosphor in culture using normal human mesothelial cells. The results showed that glass phosphor exhibited no significant changes in the cell morphology and the numbers of proliferating cells (Figure 7), suggesting that placing the clip containing the glass phosphor in the stomach for several hours has no toxic side effects on the normal human tissue.

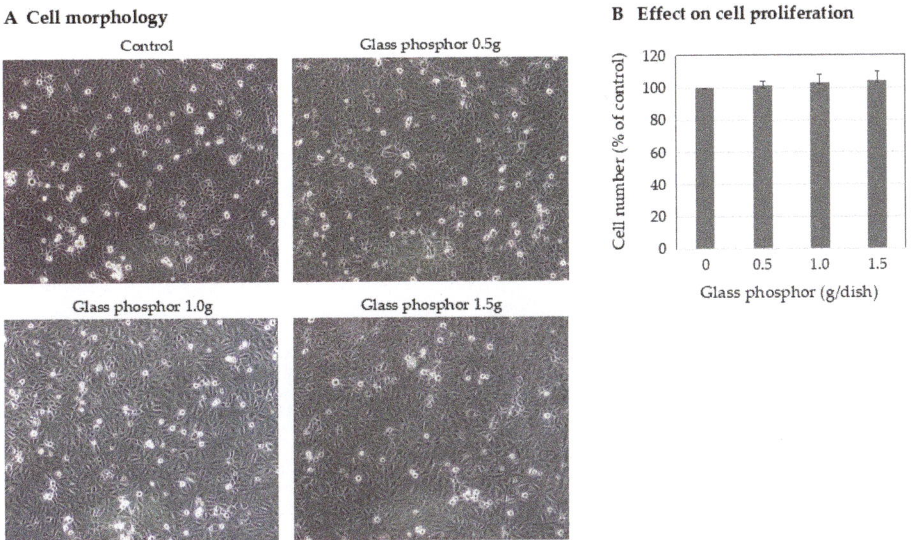

Figure 7. Effects of glass phosphor on the normal human mesothelial cells in culture. The cells were cultured in dishes containing glass phosphor (weights: 0.5, 1.0, and 1.5 g) for 72 h. (**A**) No significant change in the cell morphology was observed. (**B**) There was statistically no significant difference (NS) in the cell proliferation (by *t*-test). Bar = standard deviation.

4. Discussion

During laparoscopic surgery, the conventional marking method of using an endoscopic injection of charcoal ink around the tumor tissue makes it considerably difficult to identify the exact location of gastric tumors intraperitoneally because of the diffusion of the ink. This marking method generates a risk of over-resection with unnecessarily wide margins containing tumor-free tissue. To overcome this problem, two combination methods have recently been reported to precisely determine the resection margin. One method is intraoperative endoscopy in combination with laparoscopic surgery [10], and the other is preoperative endoscopic clipping in combination with intraoperative radiography [11]. However, these two methods require an additional diagnostic apparatus and manpower. In contrast, the new fluorescent clip with glass phosphor proposed in the present study emits NIR light (around 1000 nm), which passes through the gastric wall and can be detected by fluorescent laparoscopy alone. This new method does not need further equipment and/or manpower but has the potential to accurately identify gastric tumors during laparoscopic surgery. To the best of our knowledge, this is the first report of a combined NIR light-emitting clip and laparoscopic detection system with the ability to visualize a precise tumor location during laparoscopic surgery of gastric cancer.

Our fluorescent clip with glass phosphor has the following two unique characteristics: (1) The wavelength emitted by the glass phosphor is known to be dependent on the composition of the doping elements. We previously fabricated Yb^{3+} and Nd^{3+} co-doped Bi_2O_3-B_2O-based glass as a new NIR fluorescent light source [12]. As the luminescence of Yb^{3+} and Nd^{3+} is located in the wavelength region around 1000 nm after excitation by light (808 nm) [13], we applied this glass phosphor in the clip as a fluorescence light source that can pass through stomach wall and enable the identification of the location of the tumor from outside the stomach. (2) The fluorescence intensity of emitted NIR light from glass phosphor was substantially influenced by the three-dimensional geometrical structure of the glass phosphor. Kasugai et al. previously reported that light extraction efficiency was dependent on the surface structure of light-emitting diodes, such as the moth-eye structure [14]. Chan et al. also recently reported that the moth-eye structure inspired anti-reflective surfaces for improved IR optical systems [15]. Therefore, we examined the influence of the structure of glass phosphor and found that among various structures, the square pyramid array in two or four faces exhibited a significantly greater light intensity output than the simple rectangular structure. In this study, therefore, we used the clip with Yb^{3+} and Nd^{3+} co-doped glass phosphor in the aforementioned pyramid array structure.

The important finding of the present study was the acquisition of preclinical proof of the utility of the proposed fluorescent clip and laparoscopic detection system using two ex vivo experimental models, including resected pig and human stomach. The fluorescence intensity emitted from the clip through the stomach wall increased depending on the excitation light power (500–1500 mW) in both experiments. In the ex vivo experiment using a resected human stomach, the thickness of the stomach was 10 mm on average, which was thicker than the pig stomach (3–10 mm). In addition, the freshly resected human stomach tissue still contained a significant amount of blood unlike the commercially available pig stomach. Even in this severe condition, a strong fluorescence intensity was successfully detected from outside of the human stomach when the excitation light and rigid scope were close to the stomach wall and fixed at distances of about 6 cm from the outside of the stomach, suggesting the possibility that this combined NIR light-emitting clip and laparoscopic detection system is applicable to laparoscopic surgery in clinical settings.

In conclusion, there are still some limitations to clearly detecting NIR fluorescence from a greater distance (6–8 cm) outside of the stomach. Further improvement of the sensitivity of the laparoscopic NIR light detection system, especially by increasing the transmission efficiency of the NIR light through the rigid laparoscope, is needed. Nevertheless, the present findings strongly suggest that the proposed clip with glass phosphor and the laparoscopic detection system could be useful and practical diagnostic tools for navigation during laparoscopic gastric surgery in the clinical setting in the near future.

Author Contributions: Investigation: S.A.I., H.N., A.I. and K.M.; Resources: M.O.; Supervision: K.M., J.H. and S.F.

Funding: This research was funded by a Grant-in-Aid for Priority Research Project from Knowledge Hub Aichi, Japan.

Acknowledgments: We thank M. Yoshimura for her expert technical assistance. We thank Kelly Zammit, from Edanz Editing (www.edanzediting.com/ac), for editing a draft of this manuscript.

Conflicts of Interest: The authors declare no conflicts of interests.

References

1. Ferlay, J.; Soerjomataram, I.; Dikishit, R.; Eser, S.; Matters, C.; Rebelo, M.; Parkin, D.M.; Forman, D.; Bray, F. Cancer Incidence and mortality worldwide: Sources, methods and major patterns in Globocan. *Int. J. Cancer* **2015**, *136*, E359–E386. [CrossRef] [PubMed]

2. Nagini, S. Carcinoma of the stomach: A review of epidemiology, pathogenesis, molecular genetics and chemoprevention. *World J. Gastrointest. Oncol.* **2012**, *4*, 156–169. [CrossRef]

3. Kitano, S.; Shiraishi, N.; Uyama, I.; Sugihara, K.; Tanigawa, N. A multicenter study on oncologic outcome of laparoscopic gastrectomy for early cancer in Japan. *Ann. Surg.* **2007**, *245*, 68–72. [CrossRef]

4. Kim, H.H.; Hyung, W.J.; Cho, G.S.; Kim, M.C.; Han, S.U.; Kim, W.; Ryu, S.W.; Lee, H.J.; Song, K.Y. Morbidity and mortality of laparoscopic gastrectomy versus open gastrectomy for gastric cancer: An interim report-a phase III multicenter, prospective, randomized trial (KLASS trial). *Ann. Surg.* **2010**, *251*, 417–420. [CrossRef]

5. Katai, H.; Sasako, M.; Fukuda, H.; Nakamura, K.; Hiki, N.; Saka, M.; Yamaue, H.; Yoshikawa, T.; Kojima, K. Safety and feasibility of laparoscopy-assisted distal gastrectomy with suprapancreatic nodal dissection for clinical stage I gastric cancer: A multicenter phase II trial (JCOG 0703). *Gastric Cancer* **2010**, *13*, 238–244. [CrossRef]

6. Kitamura, K.; Takahashi, T.; Yamaguchi, T.; Yamane, T.; Hagiwara, A.; Oyama, T. Identification, by activated carbon injection, of cancer lesion during laparoscopic surgery. *Lancet* **1994**, *343*, 789. [CrossRef]

7. Tokuhara, T.; Nakata, E.; Tenjo, T.; Kawai, I.; Satoi, S.; Inoue, K.; Araki, M.; Ueda, H.; Higashi, C. A novel option for preoperative endoscopic marking with India ink in totally laparoscopic distal gastrectomy for gastric cancer: A useful technique considering the morphological characteristics of the stomach. *Mol. Clin. Oncol.* **2017**, *6*, 483–486. [CrossRef] [PubMed]

8. Anderson, R.R.; Parrish, J.A. The optics of human skin. *J. Investig. Dermatol.* **1981**, *77*, 13–19. [CrossRef]

9. Kakiuchi, T.; Takahara, T.; Kasugai, Y.; Arita, K.; Yoshida, N.; Karube, K.; Suguro, M.; Matsuo, K.; Nakanishi, H.; Kiyono, T.; et al. Modeling mesothelioma utilizing human mesothelial cells reveals involvement of phospholipase-C beta 4 in YAP-active mesothelioma cell proliferation. *Carcinogenesis* **2016**, bgw084. [CrossRef] [PubMed]

10. Kawakatsu, S.; Ohashi, M.; Hiki, N.; Nunobe, S.; Nagino, M.; Sano, T. Use of endoscopy to determine the resection margin during laparoscopic gastrectomy for cancer. *BJS* **2017**, *104*, 1829–1836. [CrossRef]

11. Chung, J.W.; Seo, K.W.; Jung, K.; Park, M.I.; Kim, S.E.; Park, S.J.; Lee, S.H.; Shin, Y.M. A promising method for tumor localization during total laparoscopic distal gastrectomy: Preoperative endoscopic clipping based on negative biopsy and selective intraoperative radiography findings. *J. Gastric Cancer.* **2017**, *17*, 220–227. [CrossRef] [PubMed]

12. Fuchi, S.; Sakano, A.; Takeda, Y. Wideband Infrared Emission from Yb^{3+} and Nd^{3+} Doped Bi_2O_3-B_2O_3 Glass Phosphor for an Optical Coherence Tomography Light Source. *Jpn. J. Appl. Phys.* **2008**, *47*, 7932–7935. [CrossRef]

13. Fuchi, S.; Sakano, A.; Mizutani, R.; Takeda, Y. High Power and High Resolution Near-Infrared Light Source for Optical Coherence Tomography Using Glass Phosphor and Light Emitting Diode. *Appl. Phys. Express* **2009**, *2*, 032102. [CrossRef]

14. Sakurai, H.; Kondo, T.; Suzuki, A.; Kitano, T.; Mori, M.; Iwaya, M.; Takeuchi, T.; Kamiyama, S.; Akasaki, I. Fabrication of high-efficiency LED using moth-eye structure. *Proc. SPIE* **2011**, *7939*. [CrossRef]

15. Chan, L.; Morse, D.; Gordon, M. Moth eye-inspired anti-reflective surfaces for improved IR optical systems & visible LEDs fabricated with colloidal lithography and etching. *Bioinspir. Biomim.* **2018**, *13*, 041001. [PubMed]

 *micromachines*

Article

Optical Properties of Au-Based and Pt-Based Alloys for Infrared Device Applications: A Combined First Principle and Electromagnetic Simulation Study

Min-Hsueh Chiu [1], Jia-Han Li [1,*] and Tadaaki Nagao [2,3,*]

1   Department of Engineering Science and Ocean Engineering, National Taiwan University, No. 1, Sec. 4, Roosevelt Rd., Taipei 10607, Taiwan; peter810601@gmail.com
2   International Center for Materials Nanoarchitectonics (WPI-MANA), National Institute for Materials Science (NIMS), 1-1 Namiki, Tsukuba, Ibaraki 305-0044, Japan
3   Department of Condensed Matter Physics Graduate School of Science, Hokkaido University, Kita-10 Nishi-8 Kita-ku, Sapporo 060-0810, Japan
*   Correspondence: jiahan@ntu.edu.tw (J.-H.L.); NAGAO.Tadaaki@nims.go.jp (T.N.)

Received: 28 December 2018; Accepted: 15 January 2019; Published: 20 January 2019

Abstract: Due to the rapid progress in MEMS-based infrared emitters and sensors, strong demand exists for suitable plasmonic materials for such microdevices. We examine the possibility of achieving this goal by alloying other metals with the noble metals Au and Pt, which have some drawbacks, such as low melting point, structural instability, and high costs. The six different metals (Ir, Mo, Ni, Pb, Ta, and W) which possess good properties for heat resistance, stability, and magnetism are mixed with noble metals to improve the properties. The optical properties are calculated by density functional theory and they are used for further investigations of the optical responses of alloy nanorods. The results show that the studied alloy nanorods have wavelength selective properties and can be useful for infrared devices and systems.

Keywords: alloy; infrared device; gold; platinum; permittivity; quality factor; first principle

---

1. Introduction

According to Planck's law, the thermal radiation of an ideal black body depends only on its temperature. However, in general, the thermal emissions from realistic objects differ substantially depending on their dielectric response and surface morphologies. Theoretical study and engineering designs of thermal emission devices are not only important in thermal emitters and infrared sensors but also useful for MEMS-based systems and micromachines. Therefore, some researchers are aiming to artificially tailor the emission spectrum by selecting the constituent materials and rationally designing their surface nanostructures to utilize thermal energy and heat management in a more efficient manner. For example, Dao et al. [1] demonstrated mid-infrared metamaterial perfect absorbers by using Al-Al$_2$O$_3$-Al tri-layers structure, which an exhibited excellent absorptivity of 98% and a flexible wavelength selective response in the infrared region tuned by the disk diameter. They also demonstrated that the sharpness of the emission peak of the device, which only uses inexpensive industrial material, was comparable to that of absorbers composed of gold or silver. Yang et al. [2] applied four elemental metals (Al, Au, Mo, W) to realize a Tamm plasmon polariton to achieve a narrow band thermal emission, which may efficiently confine the energy and result in a sharp peak in absorption spectrum in the infrared region. Yokoyama et al. [3] achieved high emission intensity by developing a thermal emitter operative at very high temperature of 1000 °C using refractive molybdenum and aluminum oxide in a metal–insulator–metal structure. Both reported the wavelength selective characteristics by adequately selecting the material and structural parameter, to achieve higher device performance in the infrared region. Regarding the operation temperature, it is an important factor which may cause the degradation of material and damage to the devices. Therefore, exploring new refractory materials is expected to bring about greater potential to work in high-temperature environments. TiN thin film has suitable plasmonic properties similar to Au as well as a substantially high thermal stability up to about 500 °C [4]. Kaur et al. [5] combined TiN nanoparticles and transparent ceramic microfiber wools to efficiently desalinate water by the broadband photothermal heat generation by absorbing ultraviolet to near-infrared light. Sugavaneshwar et al. [6] demonstrated high-quality TiN film on a flexible polymer film at room temperature, and illustrated the better and swifter conversion from light to heat energy in the infrared region.

On the other hand, due to the strong transmission and low hazard of infrared light to the living body, it is used for non-destructive sensing applications. For example, surface-enhanced Raman spectroscopy (SERS) is a technique widely used detection and sensing microdevices. With localized surface plasmons, Bansal et al. [7] simulated the effects of scattering efficiencies on the different shapes of Ag-Cu alloy, such as sphere, cube, and nanobar with an effective radius 50 nm. Dodson et al. [8] adjusted the gap between bowties to be 10 nm to pursue an extremely high enhancement factor. The effects of amplifying the electric field can be utilized in a sparse solution. The object in sparse solution can be concentrated with magnetic nanoparticles by applying the magnetic field, which reinforces the sensing efficiency. Brullot et al. [9] simulated and discussed the relations between the structure, size, and optical properties of magnetite-core gold-shell nanoparticles. They developed the comprehensive mechanism for designing the core-shell nanoparticles in the near-infrared light region. Kwizera et al. [10] successfully synthesized different shapes (sphere, popcorn, and star) and sizes of iron-oxide-core and gold-shell nanoparticles. The optical and magnetic properties of nanoparticles that may be exploited in different applications have been demonstrated. Liu et al. [11] synthesized urchin-like Ag-coated Ni nanoparticles. The rhodamine-6G sparse solution can be easily detected due to the large amount of hot spots. It has also been shown that such nanoparticles are quickly and simply manipulated by a magnetic field due to their nickel cores. Zhang et al. [12] applied nanoparticles in realistic applications and showed the differentiation of three different bacteria using Fe_3O_4-Au core-shell nanoparticles, and the signal was about 60 times higher than that obtained without a magnetic field.

Most bio-molecules vibrate in the infrared light, hence, infrared light is widely used in bio-sensing applications. Also, plasmonic nanoparticles surrounded by a catalyst are believed to improve the activities for the chemical transformations, e.g., by decreasing the chemical potential and increasing the number of free electrons. Both procedures increase the efficiency of decomposition reactions. Song et al. [13] provided an exhaustive investigation of the mechanism by which the photocatalytic reduction of carbon dioxide is achieved. The results revealed that the CO_2 decomposition efficiency of Au-Pt-SiO_2 nanoparticles was about twice as high compared to Pt-SiO_2 nanoparticles under the light intensity of 0.6 (W/cm^2); the efficiency of Au-Pt-SiO_2 nanoparticles was found to be better than that of Au-SiO_2 and Pt-SiO_2 nanoparticles. Bora et al. [14] discussed the surface temperature of Au-ZnO nanorods under resonant excitation up to about 300 °C, which gave an apparent quantum yield (for the degradation of methylene blue) about 6 times higher than that achieved by bare ZnO nanorods. Mukherjee et al. [15] showed that the H_2 dissociation can be achieved at room temperature due to the surface plasmon-induced hot electrons at the Au nanoparticle surface, as H_2 has an extremely high dissociation energy (4.51 eV).

The operation frequency is one of the most important factors influencing device efficiency. Some researchers have tuned the structure parameters of nanoparticles, as discussed above, to achieve the desired result. Instead of changing the morphology of the microstructures or nanostructures, one could focus on the optical properties of the selected materials, which are key parameters that strongly affect the plasmonic responses of materials. Different classes of materials have been explored, and they were found to depend on the operating wavelength of the applications. Some materials give a relatively low imaginary part of the permittivity as compared to metallic layers, such as graphene [16] or indium tin oxide [17,18]. Although graphene has been employed as a plasmonic material and has a negative real part of the permittivity under suitable operating conditions in the infrared light region, its magnitude of the real part of the permittivity is not so large and it gives a lower electric field enhancement on the surface of materials. To mitigate the loss of metallic materials (low imaginary part of permittivity) and maintain a relatively strong electric field enhancement (high magnitude of real part of permittivity), we propose a method of adding other components to obtain a hybrid alloy. The operating wavelength of a material can be adjusted with different atomic ratios for different applications. Silva et al. [19] gave the experimental and analytic data of AuCu alloy, which illustrated that the AuCu alloy possesses superior plasmonic properties from 500 nm to 690 nm. Keast et al. [20] simulated a series of Au-based intermetallic materials (Al, Cd, Mg, Pd, Pt, Sn, Zn, Zr) and discussed the localized surface plasmon properties of the materials. Gong et al. [21] synthesized alloys combining two out of Au, Ag, and Cu, which give a tunable wavelength characteristic in the visible light region. The authors also mentioned that the permittivity of the alloys cannot be approximated by a simple linear combination of the dielectric function from the different metals making up the composition of the material. However, most of the strong plasmonic effects of noble metals occur in the visible region, which is dominated by interband transition. Blaber et al. [22] also stated that the partially occupied d bands of the alloy may affect the efficiency of its plasmonic effects, but the dopants may disrupt the low energy transition, which would in turn increase the plasmonic quality. Hence, the alloy may suitable for applications operating in the infrared region.

In this work, we explore the materials that will be potentially operated in the infrared region, and discuss Au-based and Pt-based noble metal alloys mixed with other six metals (Ir, Mo, Ni, Pb, Ta, W). Au and Pt are widely used in the fields of plasmonics and catalysis. The permittivity of the noble metals supports a very good quality surface plasmon in the visible to infrared regions, and they are stable in normal environments. However, Au has a relatively low melting point (1377 K), while Pt has a slightly higher melting point (2041 K) but a rather high loss compared to Au. It is predicted that the melting point will increase when the alloy is combined with refractory metals such as Ir, Mo, Ta, and W, whose melting points are 2739, 2896, 3290, and 3698 K, respectively. It is shown that the tensile strength of the alloys increases with the inclusion of Ir in PtIr [23]. Moreover, Ni-based alloys may also exhibit magnetic properties due to their ferromagnetic characteristic, and thus they will be useful in nano-bio applications. It is also predicted that the thermal stability of alloys will increase with the introduction of Ni. In addition to the transition metal, the poor metal is also discussed. Compared to the transition metals, the poor metals have a higher electronegativity, a lower melting point, and may be used in organic and biomolecule applications. Pb is one of the poor metals that is reported to show plasmonic properties in nanoparticles in the ultraviolent region [24]. In this work, we focus on the infrared optical properties of these unexplored alloys and discuss their potential to be used in infrared devices and systems. To investigate the optical responses of some practical nano objects, nanorods made of pure metals and different alloy materials are examined by finite-different time-domain simulations. These nanorods can be applied in high-temperature environments, localized surface plasmon polaritons, and catalysts. It should be noted that some of the alloys were found to have better plasmonic properties or magnetic properties than pure elemental metals, indicating the effectiveness of the alloying strategy for exploring high-performance infrared devices.

2. Materials and Methods

2.1. Materials

In this paper, we consider bi-metal alloys which are a combination of one noble metal and another metallic element. For the studied bi-metal alloys, the noble metal is Au or Pt and the other metallic element is Ir, Mo, Ni, Pb, Ta, or W in our simulations. The simulated materials include 12 types of bi-metal alloys. The structures of the pure metallic materials in normal conditions are listed in Table 1.

Table 1. Space group of the elements Au, Pt, Ni, Ir, Pb, Mo, Ta, and W.

Space Group	F3m3	I3m3
Composition	Au, Pt, Ir, Ni, Pb	Mo, Ta, W

The crystal structure and electronic structure of the alloys are important factors influencing the alloys' material properties. Theoretically, it is predicted that the crystal structure of a bi-metal alloy remains the same as its constituted elements if the constituted elements have the same crystal structure. Boschow et al. [25] showed that $Au_{92}Ni_8$ exhibits the same space group, F3m3, as the constituent elements. However, some researchers have showed that the crystal structure can be reshaped and transformed into other crystal structure with an equal ratio of compositions and very different Bohr radii of elements. For example, Leroux [26] recorded the phase diagram of PtNi. It was shown that the crystal structure of PtNi alloy with an atomic ratio of around 0.5 is quite different from the crystal structures of Pt-rich or Ni-rich alloys. More complicated phase diagrams of bi-metal alloys can be found for constituted elements with two different structures. Ocken et al. [27] illustrated the phase diagram of PtMo from body-centered cubic (BCC) Mo to face-centered cubic (FCC) Pt. From the Inorganic Crystal Structure Database (ICSD [28]), it was shown that that AuTa alloy is of the F3m3 space group for $Au_{0.9}Ta_{0.1}$ but has a CsCl structure for $Au_{0.5}Ta_{0.5}$. For systematic discussions, we simulated the selected 12 alloys with I3m3 and F3m3 structures which may appear in nature. The snapshots of the $Pt_xMo_{(1-x)}$ F3m3 and I3m3 structures with different morphologies are shown in Figures 1 and 2, respectively.

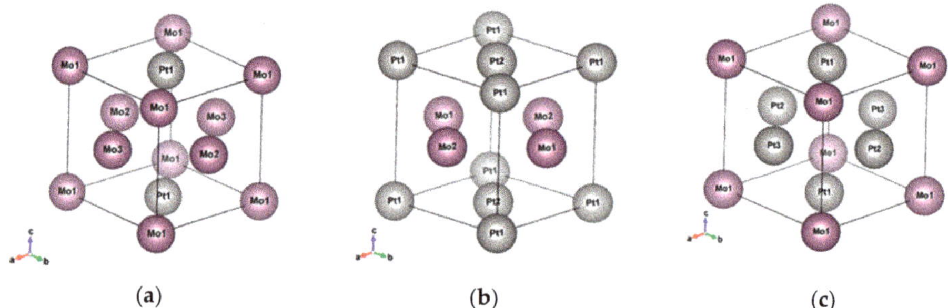

(a)   (b)   (c)

Figure 1. Snapshots of the $Pt_xMo_{(1-x)}$ F3m3 structures with different morphologies: (a) $Pt_{0.25}Mo_{0.75}$, (b) $Pt_{0.5}Mo_{0.5}$_flat, and (c) $Pt_{0.75}Mo_{0.25}$.

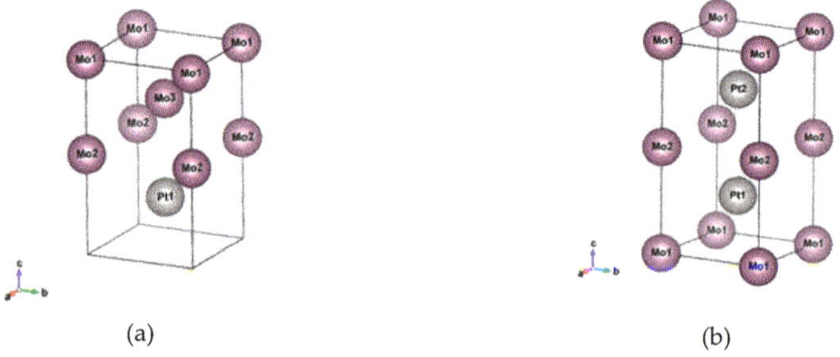

(a)   (b)

Figure 2. *Cont.*

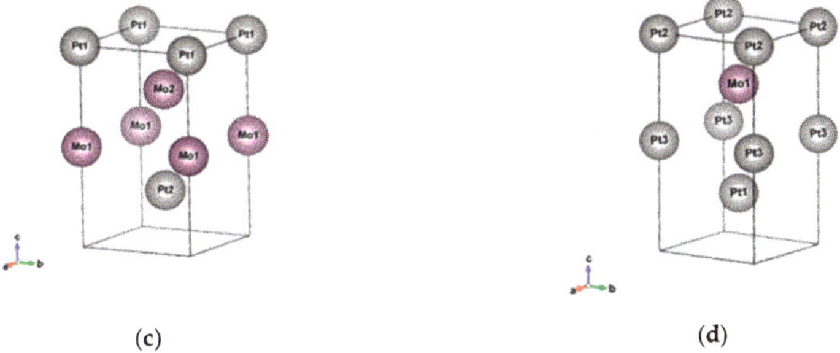

(c)   (d)

Figure 2. Snapshots of the $Pt_xMo_{(1-x)}$ I3m3 structures with different morphologies: (**a**) $Pt_{0.25}Mo_{0.75}$, (**b**) $Pt_{0.5}Mo_{0.5}$_mix, (**c**) $Pt_{0.5}Mo_{0.5}$_upper, and (**d**) $Pt_{0.75}Mo_{0.25}$.

2.2. Methods

The results of the electronic structures of alloys were calculated using a plane wave basis with a cut-off energy of 450 eV and the generalized gradient approximation (GGA) of the Perdew–Burke –Ernzerhof (PBE) [29] exchange correlation functional. The total energy was minimized by the quasi-Newton algorithm with a limitation of 0.0001 eV/Å. Spin polarization was considered for nickel-containing alloys. The ground state was calculated with $10 \times 10 \times 10$ k-points in momentum space. For optical permittivity calculation, a k-points grid of $50 \times 50 \times 50$ was chosen to ensure sufficient description for Brillouin zone. All k- points divisions are according to the Monkhorst–Pack method [30].

The finite-difference time-domain (FDTD, Lumericlal) calculations were performed to predict the optical responses of the bi-metal alloy nanorods. The wavelength-dependent dielectric response of alloys was calculated using density functional theory (DFT), and then used to evaluate the optical responses of a single nanorod. The light source was set as the plane wave, which has a polarization parallel to the nanorod. The mesh size was set as one order of magnitude smaller than the structure size.

3. Pure Au

The permittivity of bulk gold was first calculated and compared with the experimental data [31–39] to check the accuracy of the computation procedures. The permittivity was constructed by the interband and intraband contributions. The intraband is related to the behavior of the free electrons, which can be depicted by Drude's model. The real and imaginary parts of the intraband are expressed by Equations (1) and (2):

$$\mathrm{Re}[\varepsilon^{intra}(\omega)] = 1 - \frac{\omega_p^2}{\omega^2 + \Gamma^2} \tag{1}$$

$$\mathrm{Im}[\varepsilon^{intra}(\omega)] = \frac{\Gamma\omega_p^2}{\omega^3 + \omega\Gamma^2} \tag{2}$$

where ω_p represents the intraband free electron plasma frequency and Γ is the damping frequency. In this paper, ω_p is calculated by DFT and Γ is obtained from the experimental data or high accuracy simulation.

To determine the effect of Γ, we took Au as an example. In Table 2, the damping frequency is shown to fit Drude's model according to Blaber [40], Ordal [41], and Zeman [42]. For the comparisons, the damping frequency was calculated from 0.02 to 0.3 eV, and the experimental data of imaginary and real parts of dielectric constants are shown in Figure 3a,b. DF_Im. and DF_Re. along the y-axes in Figure 3 are denoted as the imaginary and real parts of the dielectric functions. It was found that the trend of Badar's data [31] which has a fitting parameter Γ with a value of 0.0184 eV, is similar to the calculated case of Γ equal to 0.02 eV. Zeman's [42] data was fitted by Hagemann's [32] experimental results which are located between 0.05 and 0.1 eV. Although Hagemann only provided experimental data under 826 nm, the trend of this data is similar to that of our simulation data. Olmon et al. [35] discussed the relationship between the parameter Γ and the permittivity of the different morphologies of gold in detail.

Table 2. Different damping frequency of gold reported by Blaber [40], Ordal [41], and Zeman [42].

Damping Frequency (eV)	Blaber [40]	Ordal [41]	Zeman [42]
Au	0.0184	0.0267	0.0708

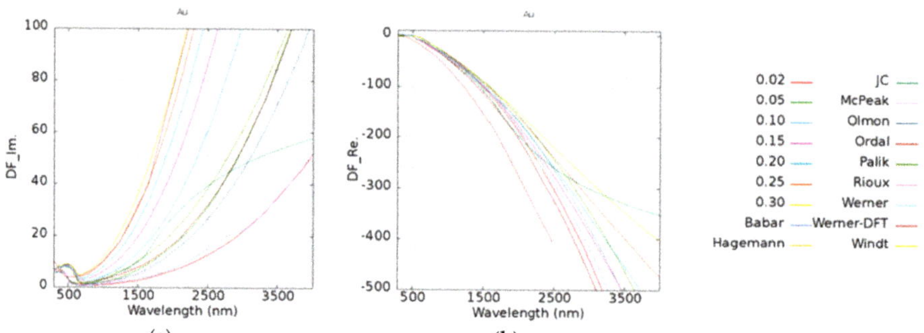

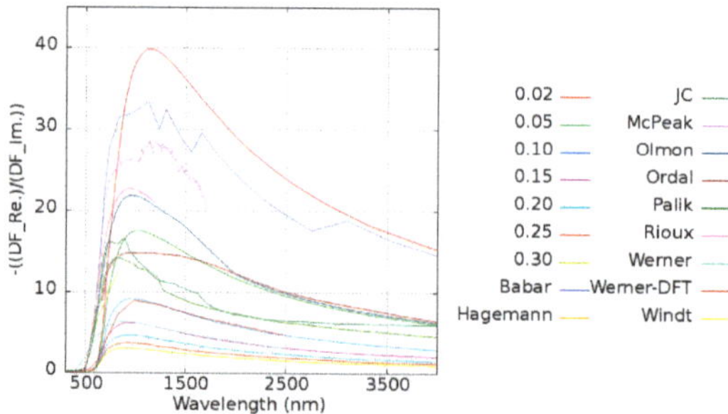

Figure 3. Testing of the damping frequency which varies from 0.02 to 0.3 in Au on (**a**) imaginary and (**b**) real parts of the permittivity. Different references [31–39] are plotted for comparison. DF_Im. and DF_Re. along the *y*-axes denote as the imaginary and real parts of the dielectric functions.

To excite the plasmons, the real part of the dielectric function should be negative to match the condition of the adjacent dielectric material. The parameter quality factor (Q) is introduced to examine the performance of plasmons, and it is defined as Equation (3).

$$Q = -\frac{\mathrm{Re}[\varepsilon(\omega)]}{\mathrm{Im}[\varepsilon(\omega)]} \tag{3}$$

For a given material, a high Q means that it has less loss and a stronger resonance property for inducing plasmons in this frequency range. The Q of pure Au is shown in Figure 4. The maximum of Q is located at around 1100 nm, which is consistent with the experimental data [31–39]. Most of the experimental data's Γ are located between 0.02 and 0.1 eV. Considering the different metals employed in this work, the damping frequency is taken as 0.1 eV for all systems except pure Au.

Figure 4. Quality factor versus wavelength for the damping frequency varying from 0.02 to 0.3 eV for Au. The quality factors calculated from the dielectric constants obtained from different references [31–39] are plotted for comparison.

4. Au-Based Alloy

The permittivity of alloys was calculated by combining the DFT calculated interband transition and Drude's term of intraband transition. To justify the calculated results, we took AuNi as an example. Jain et al. [43] reported an AuNi alloy wire generated by melting 99.99% pure nickel and gold ingot. It was predicted that the alloy atomic ratio would be 0.5 due to the vacuum system and repeated re-melting process. The quality factors of the AuNi from Jain's results were 0.575, 0.660, and 0.606 at 1000, 1500, and 2000 nm wavelengths, respectively. The trend of the quality factor was consistent with our data on $Au_{0.5}Ni_{0.5}$ (BCC_mix), which exhibited quality factors of 0.668, 0.686, and 0.682 at the same wavelengths.

To inspect the optical properties of alloys, we took $Au_xMo_{(1-x)}$ alloy as an example. The imaginary and real parts of the permittivity of $Au_xMo_{(1-x)}$ alloy are shown in Figure 5a,b, respectively. In the legend of the plots in Figure 5, the first and the last ones characterize the pure metallic elements. The number represents the atomic ratio of Au, while the text after the number denotes the structure and the morphology under the given atomic ratio condition. For example, 0.25_bcc represents $Au_{0.25}Mo_{0.75}$ with a BCC structure. In the $Au_xMo_{(1-x)}$ alloy, all cases possessed a higher imaginary part of the permittivity than pure Au and Mo above 1000 nm. In addition to intraband transition, the interband transition also has the contribution to the imaginary part of the permittivity. Here, $Au_{0.25}Mo_{0.75}$ (FCC) was taken as an example for further discussion. The permittivity contributed by the interband transition of $Au_{0.25}Mo_{0.75}$ (FCC), pure Au, and Mo is shown in Figure 5c, and the corresponding density of states (DOS) is shown in Figure 5d. The DOS for each case is normalized by the maximum value within the range of -10 to 5 eV. It was shown that the onset of interband transition is roughly consistent with the distance between the large quantity states and the Fermi level. For instance, it was shown that the first huge DOS peak below the Fermi's level is around 2.5 eV in the case of Au (Figure 5d, blue line), whereas the occurrence of interband transition is around 516 nm. The combination of Au and Mo leads to additional states, which are known as the virtual states [20] and add the possibility of photon absorptions for the higher imaginary part of the permittivity.

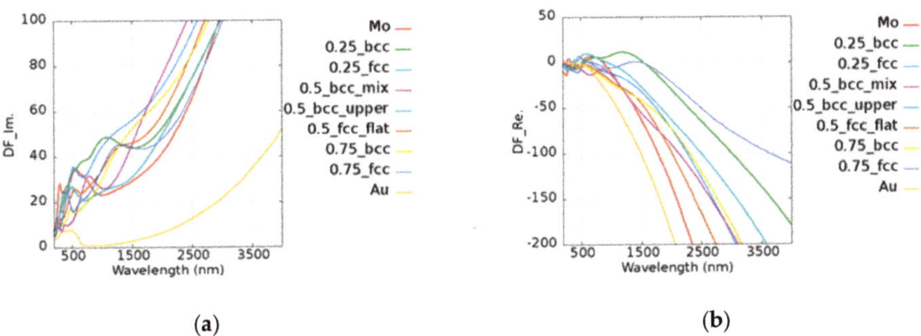

(**a**)　　　　　　　　　　(**b**)

Figure 5. *Cont.*

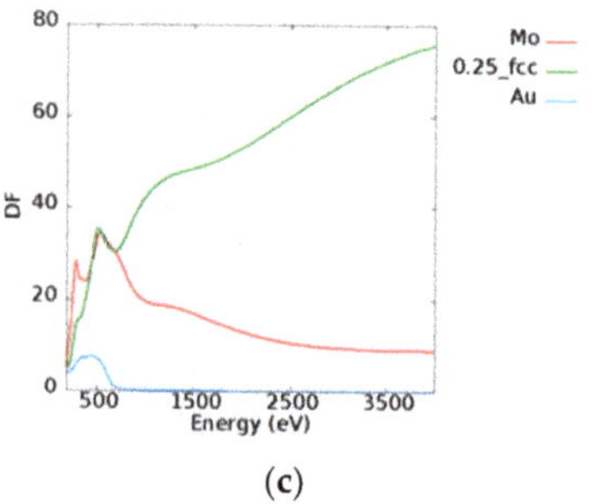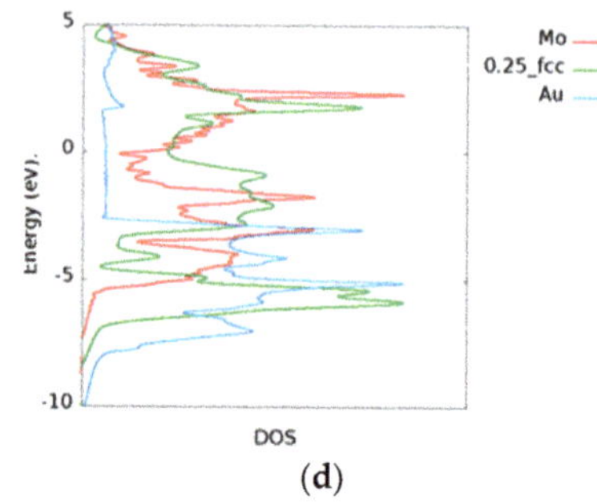

(c)   (d)

Figure 5. (**a**) Imaginary and (**b**) real parts of the permittivity of AuMo alloy. (**c**) The interband transition of the imaginary part of the permittivity and (**d**) the density of states of Mo, $Au_{0.25}Mo_{0.75}$_FCC, and Au. DF_Im. and DF_Re. along the *y*-axes denote the imaginary and real parts of the dielectric functions.

The results of Q of the Au-based alloy are shown in Figure 6. In the legend of the plots, the first and the last ones characterize the pure metallic elements. The number represents the atomic ratio ofe Au, while the text after the number denotes the structure and the morphology under the given atomic ratio condition. For example, in Figure 6a, 0.5_bcc_mix represents $Au_{0.5}Ir_{0.5}$_mix with a BCC structure. The extreme high Q of Au is not fully shown here, although it is given in Figure 4. A material which has a Q value lower than 1 may not be a good plasmonic material in such a frequency. Low-Q materials occur around the visible region in most cases, and this contributes to the relatively high real part of the permittivity and the onset of interband transition. Some cases even serve as a dielectric material; in other words, they have a positive real part of the permittivity, as shown in the Appendix A. Although some alloys possess lower imaginary parts compared to pure metal, such as AuIr and AuNi, it was found that alloys have a larger real part of the permittivity due to the reduction of the plasma frequency. This leads to a larger intraband transition, as shown Equation (1), causing the shrinking of Q. Most of the cases exhibited a smaller Q than pure metal, but there are still several instances that showed a good property of Q. For example, $Au_{0.25}Ir_{0.75}$ and $Au_{0.75}Ir_{0.25}$ had a better performance than pure metal at 600 nm and 800 nm of narrow band. The extremely low imaginary part of the permittivity of $Au_{0.5}Ni_{0.5}$ (FCC) was found even compared with other Au-based alloys and pure Ni. The real and imaginary parts of permittivities of Au-based alloy are shown in Figure A1 of Appendix A. It is not surprising that $Au_{0.5}Ni_{0.5}$ (FCC) showed a good performance above 3500–4000 nm. $Au_{0.5}W_{0.5}$ (BCC_mix) possessed a better Q performance than pure W around the near-infrared region at 1000 nm. On the other hand, manufacturing costs may be reduced by using different compositions. The prices of Au, Pt, and Ir remained in same order in 2018, being more expensive than Ni, Mo, Pb, Ta, and W. The raw material cost may be an important issue affecting the mass production of devices. The Q performance was larger than 1 in most of the Au-based alloys in the infrared region, indicating that these may be used in different applications.

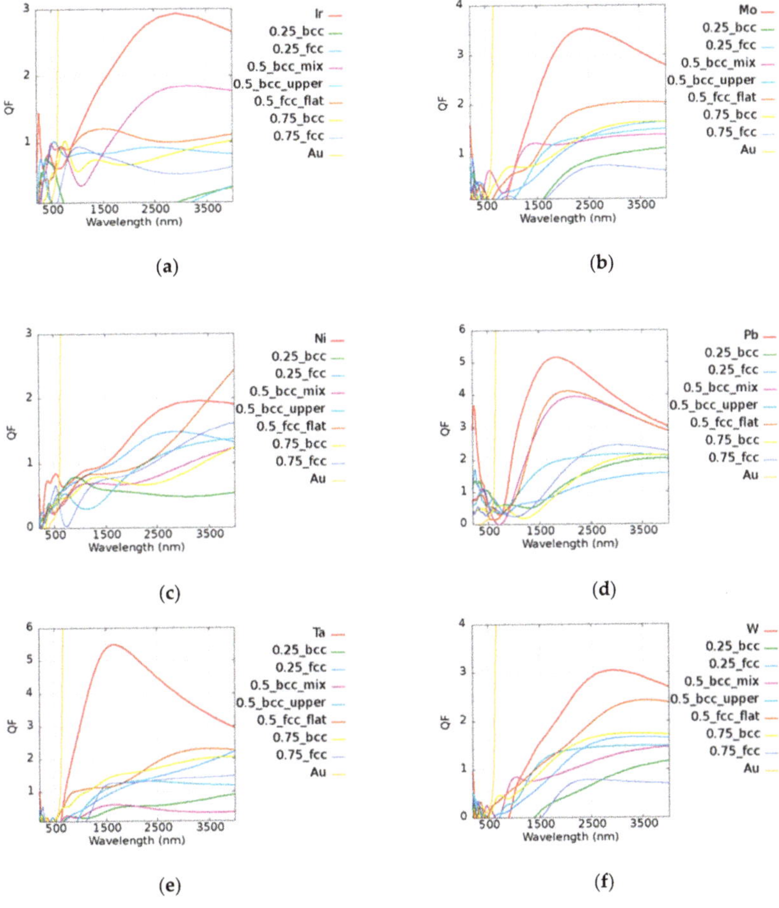

Figure 6. Quality factors of (**a**) $Au_xIr_{(1-x)}$, (**b**) $Au_xMo_{(1-x)}$, (**c**) $Au_xNi_{(1-x)}$, (**d**) $Au_xPb_{(1-x)}$, (**e**) $Au_xTa_{(1-x)}$, and (**f**) $Au_xW_{(1-x)}$ alloys.

5. Pt-Based Alloy

Unlike pure Au, Pt has a smaller Q due to its high imaginary part of the permittivity. The results of Q of Pt-based alloy are shown in Figure 7. In the legend of the plots, the first and the last ones characterize the pure metallic elements. The number represents the atomic ratio of Au, while the text after the number denotes the structure and the morphology under the given atomic ratio condition. For example, in Figure 7a, 0.75_fcc represents $Au_{0.75}Ir_{0.25}$ with an FCC structure. $Pt_{0.5}Ir_{0.5}$ (BCC_mix) and $Pt_{0.75}Ir_{0.25}$ (BCC) had stronger plasmonic responses in the visible range of 600 to 700 nm as compared to pure Pt. $Pt_{0.75}Mo_{0.25}$ (BCC_mix) exhibited a higher Q in the range of 900 to 1300 nm, which contributes to the lower loss and similar real part of pure Pt. The real and imaginary parts of permittivities of Pt-based alloy are shown in Figure A2 of Appendix A. $Pt_{0.75}Ni_{0.25}$ (BCC) could be applied in the range of 600 to 1000 nm and $Pt_{0.25}Ni_{0.75}$ (FCC) had a similar Q to that of two pure metals in the range of 2400 to 2700 nm. $Pt_{0.25}Pb_{0.75}$ (FCC) and $Pt_{0.5}Pb_{0.5}$ (BCC_upper) exhibited better results above 2750 nm compared to pure Pt. $Pt_{0.5}Ta_{0.5}$ (BCC_mix) was good at visible range, and $Pt_{0.5}Ta_{0.5}$ (BCC_upper) and $Pt_{0.5}Ta_{0.5}$ (BCC_flat) could be applied above 2700 and 2000 nm, respectively. $Pt_{0.5}Ta_{0.5}$ (BCC_mix) had a high Q in the broad band of 600 to 1400 nm. Pt was found to be a good candidate for catalysis, as mentioned in the previous discussions. It was shown that some alloys possess better performances than pure Pt. According to our simulation results, the design of the material used in devices may be adjusted in different applications in consideration of the cost and the performance.

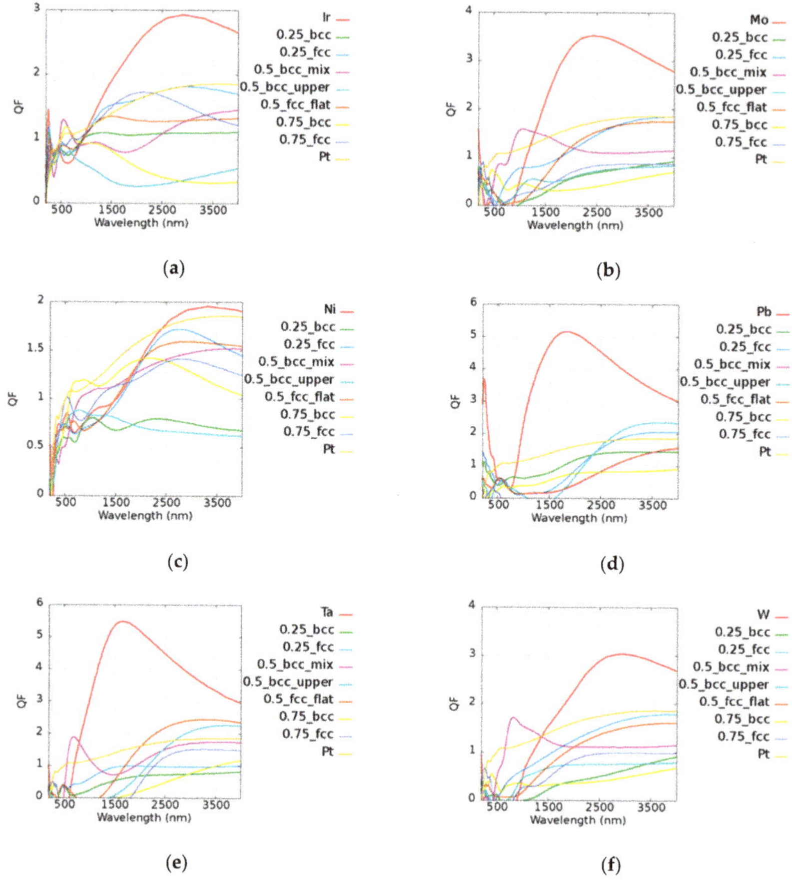

Figure 7. Quality factors of (**a**) $Pt_xIr_{(1-x)}$, (**b**) $Pt_xMo_{(1-x)}$, (**c**) $Pt_xNi_{(1-x)}$, (**d**) $Pt_xPb_{(1-x)}$, (**e**) $Pt_xTa_{(1-x)}$, and (**f**) $Pt_xW_{(1-x)}$ alloys.

6. Discussions

Most of the cases exhibited fine performances as $Q > 1$ in the infrared frequency, which indicated the good applicability of most alloys. It was found that Q was approximated to be inversely proportional to the frequency in the infrared frequency region because it is dominated by the motion of free electron according to the Drude model, as the incident photon energy is low in the infrared region.

For the applications of magnetic plasmonics, the ferromagnetic metal Ni was alloyed with Au and Pt. Our calculated optical property of $Au_{0.75}Ni_{0.25}$ has the similar trend as compared to the experimental data by using dual magnetron sputtering to deposit the metallic alloy thin film from Mcpherson et al. [44]. Lonergan et al. [45] studied the effect of different ratios of $Pt_xNi_{(1-x)}$ on benzene hydrogenation, and it was shown that Ni in alloy strengthens the bond formation which increases the activity of catalysts. The magnetic moments of $Au_xNi_{(1-x)}$ and $Pt_xNi_{(1-x)}$ alloy are listed in Table 3. As the content of Ni increases, the magnetic moment rises. However, a trade-off between plasmonic properties and magnetic properties was observed, except that FCC $Au_{0.5}Ni_{0.5}$ exhibited better characteristics than pure Ni around 3000 to 4000 nm, which may indicate its potential as a magnetic plasmonic material.

Table 3. Magnetic moments of AuNi and PtNi with different ratios and morphologies.

Magnetic Moment	0	0.25	0.5	0.75	1
$Au_xNi_{(1-x)}$	0.617	1.573 (BCC) 1.551 (FCC)	0.496 (BCC_mix) 0.960 (BCC_upper) 0.726 (FCC)	0.006 (BCC) 0.012 (FCC)	0
$Pt_xNi_{(1-x)}$	0.617	2.240 (BCC) 2.246 (FCC)	0.705 (BCC_mix) 1.370 (BCC_upper) 2.179 (FCC)	−0.001 (BCC) 1.043 (FCC)	0

It has been reported that PtMo is suitable for use in fuel cells due to its greater stability compared to that of pure Pt. Pt cathodes suffer from oxidation and a loss of active surface area. It was shown that the alloy cathode had a lighter material loss when CO was present in the fuel stream. The PtMo alloy nanoparticles were reported by Liu [46], who successfully mitigated the contamination of the anode electrocatalysts. Ehteshami et al. [47] also showed the identical trend that PtMo had a better performance than pure Pt. It is worth mentioning that PtNi has also been reported in the literature. It was shown that the bimetallic alloy is more stable than a monometallic material in some cases, making it more suitable for catalytic and thermal applications.

7. Optical Responses of Nanorods

To further investigate the optical responses of some practical nano objects, the electromagnetic fields of nanorods made of different materials were examined. In this part, we mainly focused on the effect of the material; therefore, the nanorods were simulated. The influence of the gaps between adjacent nanoraods or the density of nanorods was neglected.

First, to confirm the results of FDTD with the experimental data from Takahashi et al. [48], a gold nanorod with a length varying from 60 to 70 nm with an interval of 5 nm and a radius of 11 nm was simulated. The absorbance of gold nanorods is shown in Figure 8, revealing peaks located at 830, 850, and 950 nm for nanorod lengths of 60, 65, and 70 nm, respectively. Because a single gold nanorod was considered in this simulation, the extreme narrow peak is shown, which is the ideal case of the pure gold nanorod. It is easy to explain that the broadened peak of the experimental data [48] is simply due to the size distribution of the nanorods. On the other hand, the wavelength tunability is clearly shown, indicating that the absorbance peak increased as the nanorod length increased. The intensity was also strengthened with length, which give a quite intuitive explanation of absorbing more energy. The results were consistent with the experimental data [48], showing an absorbance peak located at 800 to 1100 nm.

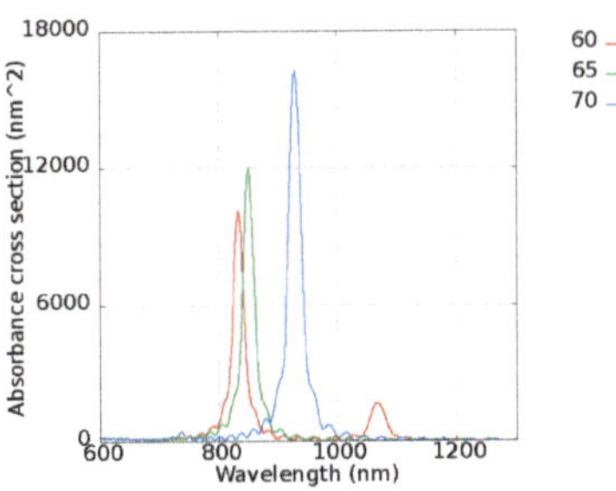

Figure 8. Absorbance cross-sections of the gold nanorods with different nanorod lengths of 60, 65, and 70 nm.

The simulated results of the alloy nanorods with a length of 100 nm and a diameter of 11 nm which have cross-section areas of approximately 1480 nm^2 are shown in Figure 9, including AuNi, AuPb, PtNi, and PtPb. Combining the two materials, the trend of the absorbance cross-section was consistent with the Q values. For example, in the AuPb case, the sequence of Q intensity above 1000 nm was Au, Pb, Au$_{0.5}$Pb$_{0.5}$ (FCC_flat), and Au$_{0.5}$Pb$_{0.5}$(BCC_mix), etc. However, the nanorod peak absorbance was confined to a small region due to the structure effect of the nanorods. The extreme narrow peak of Au is shown, illustrating its high sensitivity. On the other hand, Pt had a strong absorbance in the visible and infrared light regions, Ni showed quite a broad band from 500 to 1300 nm, and Pb possessed two main peaks at around 490 and 1100 nm. The broad band absorbance of Ni and Pb contributed to the additional absorption in the alloy above 2000 nm compared to the pure metal, which may be applied in the lower energy region. In Au-based alloy nanorods, the better plasmonic performance of alloy was demonstrated compared to pure Au in the visible region. The other better performance was shown in the infrared region around 1300 to 1400 nm in all cases compared to Au. It is worth mentioning that the relative narrow bands of Au$_{0.5}$Pb$_{0.5}$ (FCC_flat), Au$_{0.5}$Pb$_{0.5}$ (FCC_upper), and Au$_{0.5}$Pb$_{0.5}$ (BCC_mix) were found and that all peaks were situated in a similar frequency This could reduce the difficulty in synthesis by allowing some ambiguity in stoichiometry. Peak shifting was shown in most cases except for PtIr and PtNi. Due to the analogous electromagnetic response of Pt, Ir, and Ni, the alloys of these metals maintain similar operation wavelengths of 700 to 1100 nm. These alloys can be well designed by tuning their structure parameters for suitable frequency. On the contrary, the wavelength selective absorbance is shown, which may be utilized for different applications in different operation frequency ranges.

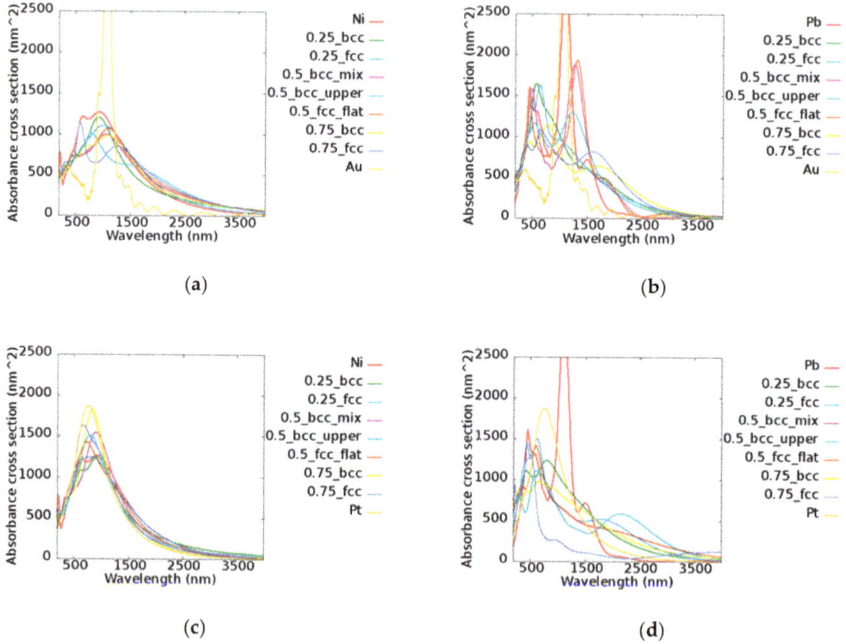

Figure 9. Absorbance cross-sections of the (**a**) AuNi, (**b**) AuPb, (**c**) PtNi, and (**d**) PtPb nanorods with a length of 100 nm and a radius of 11 nm.

The tunable absorbances of nanorods were consisted with the aspect ratio, as demonstrated in many reports [49–51]. The nanorod was lengthened to be 350 nm whereas the diameter was maintained at 11 nm. Upon increasing the aspect ratio to the value of 31.81, the absorbance cross-sections of AuNi,

AuPb, PtNi, and PtPb varied, as shown in Figure 10. The absorbance peak shifted to around 3000 nm. The absorbance cross-sections of all alloy nanorods possessed a much larger area than the area cross-sections of about 5329.34 nm^2 around the wavelength of 3000 nm. This indicates that these alloy nanorods are suitable for application in the infrared range. Effects of the wavelength tunability and associated amplification were found in different alloys. This illustrates that the compositions of two elements can be well adjusted to access different absorbance features.

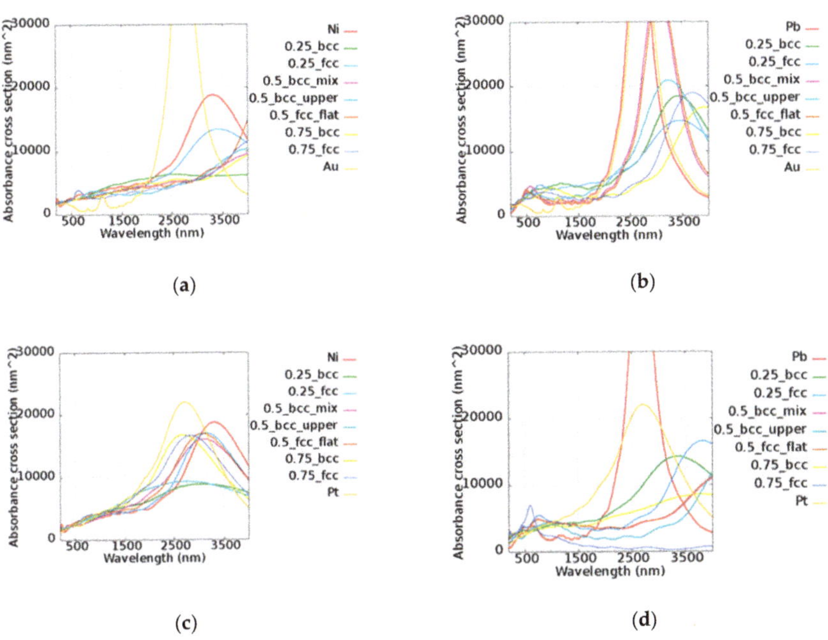

Figure 10. Absorbance cross-sections of the (**a**) AuNi, (**b**) AuPb, (**c**) PtNi, and (**d**) PtPb nanorods with a length of 350 nm and a radius of 11 nm.

8. Conclusions

In this work, the dielectric functions of Au- and Pt-based alloys were examined by DFT simulation, and then adopted in the absorbance calculations of alloy nanorods by FDTD simulations. The plasmonic performances of bulk devices were inspected by their quality factors. Most of the alloys were found to be suitable for use in the infrared region for certain compositions and would be useful for thermal emitters and infrared sensors which can be further employed in MEMS-based systems and micromachines. Furthermore, some of the alloy exhibited better performances in the infrared region compared to pure elemental metals, which is rarely seen for cases in the visible region. Pt-based alloys possessed a similar quality factor Q as that of pure metal in the infrared region, which may indicate their potential for replacing Pt in some catalytic applications. FDTD simulations for nanorods were demonstrated to examine the optical responses of these alloys. Wavelength selective properties were shown in most cases from the visible to the infrared region. The nanorods of different alloys could be applied in high-temperature environments, localized surface plasmon polaritions, and catalysts, with considering their Q performances at particular wavelengths and manufacturing cost. This work provides a method for designing suitable alloys for infrared devices.

Author Contributions: M.-H.C., J.-H.L. and T.N. gave the concepts of this paper; M.-H.C. performed the simulations; J.-H.L. analyzed the data; M.-H.C., J.-H.L. and T.N. investigated the alloy properties; M.-H.C. and J.-H.L. conceived and designed the simulations; M.-H.C., J.-H.L. and T.N. wrote the paper.

Acknowledgments: This work was partially supported by the Ministry of Science and Technology (104-2221-E-002-079-MY3 and 107-2918-I-002-020) and the NTU Project (107L7829). We are grateful to the National Center for High-Performance Computing, Taiwan, for providing us with the computation time and facilities. This work was also partially supported by JSPS KAKENHI (16F16315, JP16H06364, 16H03820) and CREST "Phase Interface Science for Highly Efficient Energy Utilization" (JPMJCR13C3) from Japan Science and Technology Agency.

Conflicts of Interest: The authors declare no conflict of interest.

Appendix A

The permittivity of the Au-based alloy and that of the Pt-based alloy are shown in Figures A1 and A2, respectively. The left-hand column is the imaginary part of the permittivity and the right-hand column is real part of the permittivity.

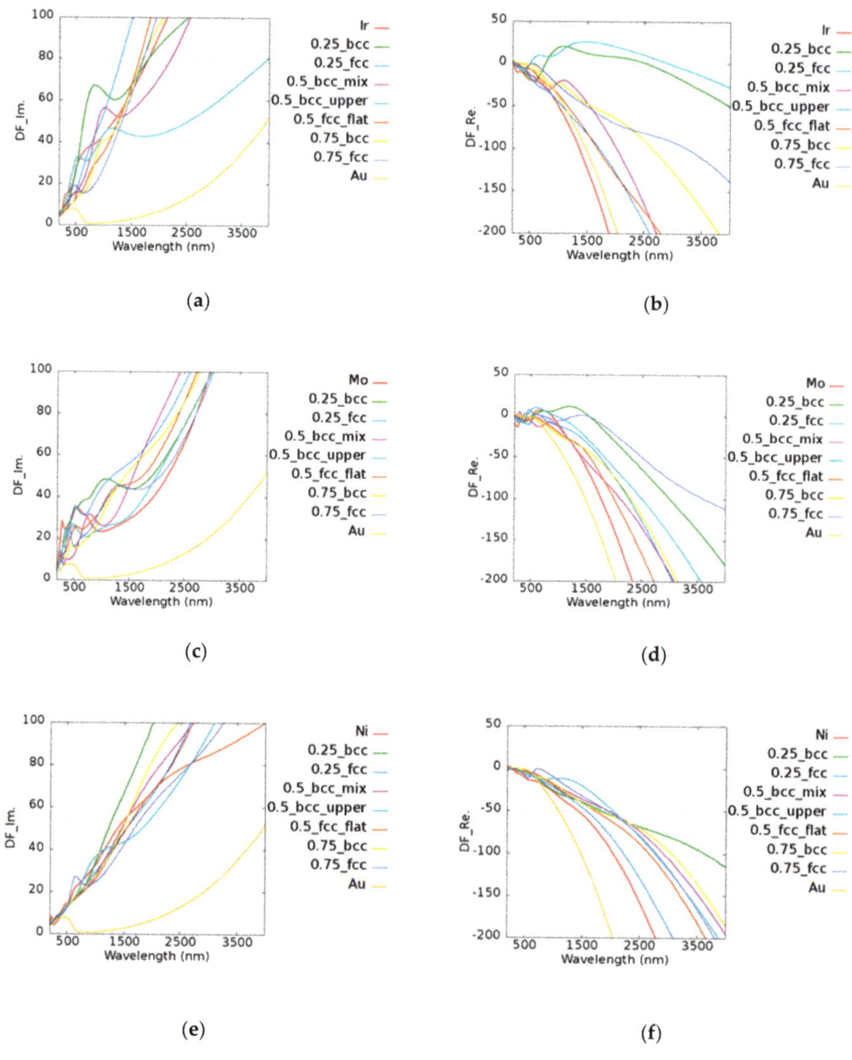

Figure A1. *Cont.*

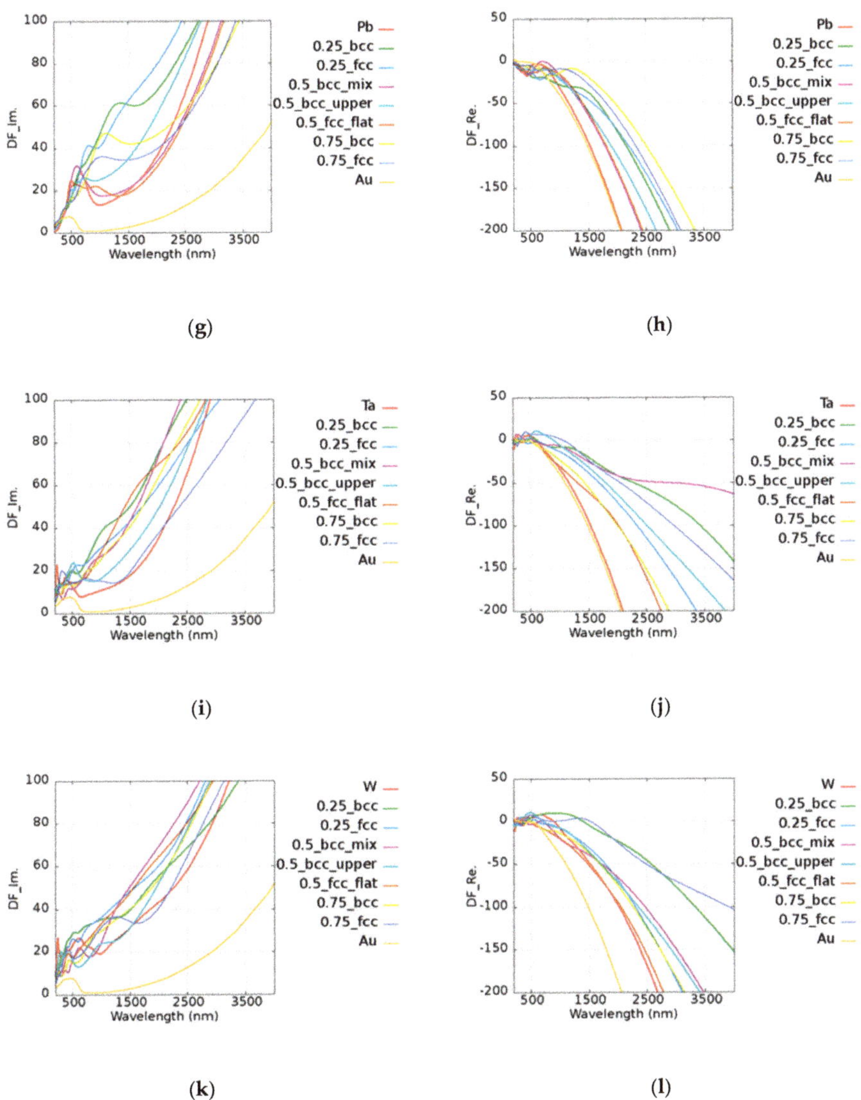

Figure A1. Imaginary and real parts of the permittivity of (**a**,**b**) AuIr, (**c**,**d**) AuMo, (**e**,**f**) AuNi, (**g**,**h**) AuPb, (**i**,**j**) AuTa, and (**k**,**l**) AuW. DF_Im. and DF_Re. along the *y*-axes denote the imaginary and real parts of the dielectric functions.

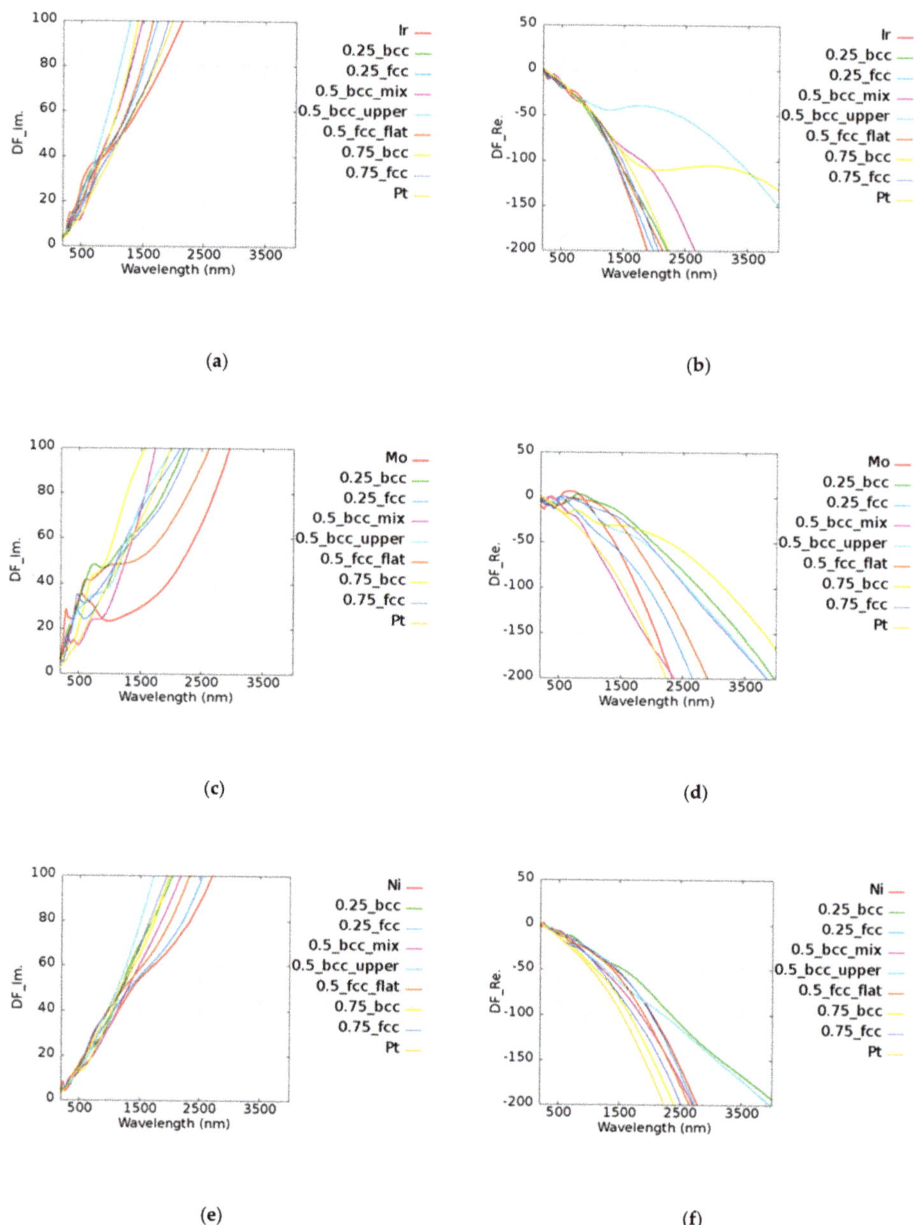

(a)

(b)

(c)

(d)

(e)

(f)

Figure A2. *Cont.*

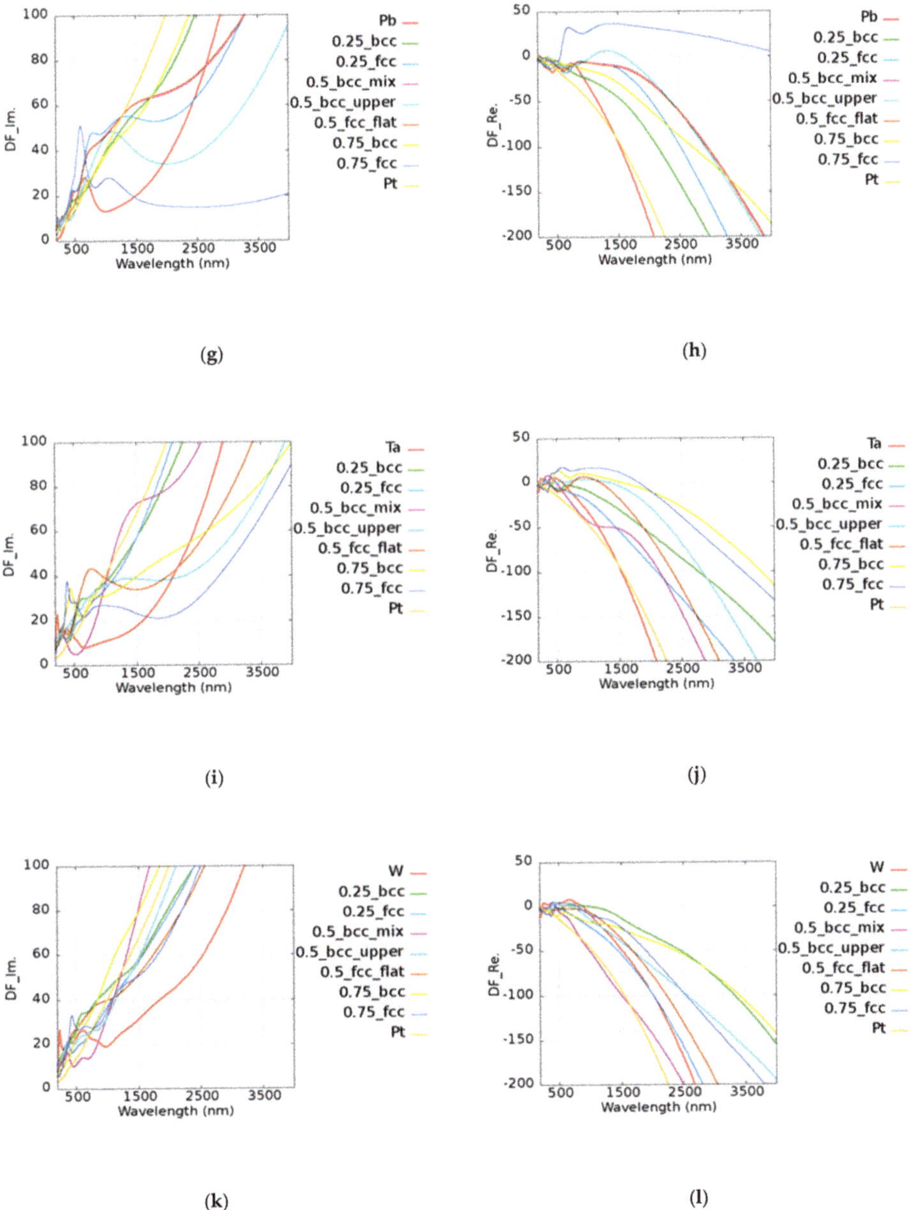

Figure A2. Imaginary and real parts of the permittivity of (**a,b**) PtIr, (**c,d**) PtMo, (**e,f**) PtNi, (**g,h**) PtPb, (**i,j**) PtTa, and (**k,l**) PtW. DF_Im. and DF_Re. along the *y*-axes denote the imaginary and real parts of the dielectric functions.

References

1. Dao, T.D.; Chen, K.; Ishii, S.; Ohi, A.; Nabatame, T.; Kitajima, M.; Nagao, T. Infrared perfect absorbers fabricated by colloidal mask etching of Al–Al$_2$O$_3$–Al trilayers. *ACS Photonics* **2015**, *2*, 964–970. [CrossRef]
2. Yang, Z.-Y.; Ishii, S.; Yokoyama, T.; Dao, T.D.; Sun, M.-G.; Pankin, P.S.; Timofeev, I.V.; Nagao, T.; Chen, K.-P. Narrowband wavelength selective thermal emitters by confined tamm plasmon polaritons. *ACS Photonics* **2017**, *4*, 2212–2219. [CrossRef]
3. Yokoyama, T.; Dao, T.D.; Chen, K.; Ishii, S.; Sugavaneshwar, R.P.; Kitajima, M.; Nagao, T. Spectrally selective mid-infrared thermal emission from molybdenum plasmonic metamaterial operated up to 1000 C. *Adv. Opt. Mater.* **2016**, *4*, 1987–1992. [CrossRef]
4. Reddy, H.; Guler, U.; Kudyshev, Z.; Kildishev, A.V.; Shalaev, V.M.; Boltasseva, A. Temperature-dependent optical properties of plasmonic titanium nitride thin films. *ACS Photonics* **2017**, *4*, 1413–1420. [CrossRef]
5. Kaur, M.; Ishii, S.; Shinde, S.L.; Nagao, T. All-ceramic microfibrous solar steam generator: TiN plasmonic nanoparticle-loaded transparent microfibers. *ACS Sustain. Chem. Eng* **2017**, *5*, 8523–8528. [CrossRef]
6. Sugavaneshwar, R.P.; Ishii, S.; Dao, T.D.; Ohi, A.; Nabatame, T.; Nagao, T. Fabrication of highly metallic TiN films by pulsed laser deposition method for plasmonic applications. *ACS Photonics* **2017**, *5*, 814–819. [CrossRef]
7. Bansal, A.; Verma, S. Optical response of noble metal alloy nanostructures. *Phys. Lett. A* **2015**, *379*, 163–169. [CrossRef]
8. Dodson, S.; Haggui, M.; Bachelot, R.; Plain, J.; Li, S.; Xiong, Q. Optimizing electromagnetic hotspots in plasmonic bowtie nanoantennae. *J. Phys. Chem. Lett.* **2013**, *4*, 496–501. [CrossRef]
9. Brullot, W.; Valev, V.K.; Verbiest, T. Magnetic-plasmonic nanoparticles for the life sciences: Calculated optical properties of hybrid structures. *Nanomed. Nanotechnol. Biol. Med.* **2012**, *8*, 559–568. [CrossRef]
10. Kwizera, E.A.; Chaffin, E.; Shen, X.; Chen, J.; Zou, Q.; Wu, Z.; Gai, Z.; Bhana, S.; O'Connor, R.; Wang, L. Size-and shape-controlled synthesis and properties of magnetic–plasmonic core–shell nanoparticles. *J. Phys. Chem. C* **2016**, *120*, 10530–10546. [CrossRef]
11. Liu, D.; Wang, X.; He, D.; Dao, T.D.; Nagao, T.; Weng, Q.; Tang, D.; Wang, X.; Tian, W.; Golberg, D. Magnetically Assembled Ni@Ag Urchin-Like Ensembles with Ultra-Sharp Tips and Numerous Gaps for SERS Applications. *Small* **2014**, *10*, 2564–2569. [CrossRef] [PubMed]
12. Zhang, L.; Xu, J.; Mi, L.; Gong, H.; Jiang, S.; Yu, Q. Multifunctional magnetic–plasmonic nanoparticles for fast concentration and sensitive detection of bacteria using SERS. *Biosens. Bioelectron.* **2012**, *31*, 130–136. [CrossRef] [PubMed]
13. Song, H.; Meng, X.; Dao, T.D.; Zhou, W.; Liu, H.; Shi, L.; Zhang, H.; Nagao, T.; Kako, T.; Ye, J. Light-Enhanced Carbon Dioxide Activation and Conversion by Effective Plasmonic Coupling Effect of Pt and Au Nanoparticles. *ACS Appl. Mater. Interfaces* **2017**, *10*, 408–416. [CrossRef] [PubMed]
14. Bora, T.; Zoepfl, D.; Dutta, J. Importance of plasmonic heating on visible light driven photocatalysis of gold nanoparticle decorated zinc oxide nanorods. *Sci. Rep.* **2016**, *6*, 26913. [CrossRef] [PubMed]
15. Mukherjee, S.; Libisch, F.; Large, N.; Neumann, O.; Brown, L.V.; Cheng, J.; Lassiter, J.B.; Carter, E.A.; Nordlander, P.; Halas, N.J. Hot electrons do the impossible: Plasmon-induced dissociation of H2 on Au. *Nano Lett.* **2012**, *13*, 240–247. [CrossRef] [PubMed]
16. Xiao, T.-H.; Cheng, Z.; Goda, K. Graphene-on-silicon hybrid plasmonic-photonic integrated circuits. *Nanotechnology* **2017**, *28*, 245201. [CrossRef] [PubMed]
17. Chen, C.; Wang, Z.; Wu, K.; Chong, H.; Xu, Z.; Ye, H. ITO–TiN–ITO Sandwiches for Near-Infrared Plasmonic Materials. *ACS Appl. Mater. Interfaces* **2018**, *10*, 14886–14893. [CrossRef]
18. Franzen, S. Surface plasmon polaritons and screened plasma absorption in indium tin oxide compared to silver and gold. *J. Phys. Chem. C* **2008**, *112*, 6027–6032. [CrossRef]
19. De Silva, K.; Gentle, A.; Arnold, M.; Keast, V.; Cortie, M. Dielectric function and its predicted effect on localized plasmon resonances of equiatomic Au–Cu. *J. Phys. D Appl. Phys.* **2015**, *48*, 215304. [CrossRef]
20. Keast, V.J.; Barnett, R.L.; Cortie, M. First principles calculations of the optical and plasmonic response of Au alloys and intermetallic compounds. *J. Phys. Condens. Matter* **2014**, *26*, 305501. [CrossRef]
21. Gong, C.; Kaplan, A.; Benson, Z.A.; Baker, D.R.; McClure, J.P.; Rocha, A.R.; Leite, M.S. Band Structure Engineering by Alloying for Photonics. *Adv. Opt. Mater.* **2018**, 1800218. [CrossRef]

22. Blaber, M.G.; Arnold, M.D.; Ford, M.J. A review of the optical properties of alloys and intermetallics for plasmonics. *J. Phys. Condens. Matter* **2010**, *22*, 143201. [CrossRef] [PubMed]

23. Rakhtsaum, G. Platinum alloys: A selective review of the available literature. *Platin. Met. Rev.* **2013**, *57*, 202–213. [CrossRef]

24. McMahon, J.M.; Schatz, G.C.; Gray, S.K. Plasmonics in the ultraviolet with the poor metals Al, Ga, In, Sn, Tl, Pb, and Bi. *Phys. Chem. Chem. Phys.* **2013**, *15*, 5415–5423. [CrossRef] [PubMed]

25. Buschow, K.V.; Van Engen, P.; Jongebreur, R. Magneto-optical properties of metallic ferromagnetic materials. *J. Magn. Magn. Mater.* **1983**, *38*, 1–22. [CrossRef]

26. Leroux, C.; Cadeville, M.; Pierron-Bohnes, V.; Inden, G.; Hinz, F. Comparative investigation of structural and transport properties of L10 NiPt and CoPt phases; the role of magnetism. *J. Phys. F Met. Phys.* **1988**, *18*, 2033. [CrossRef]

27. Ocken, H.; Van Vucht, J. Phase equilibria and superconductivity in the molybdenum-platinum system. *J. Less Common Met.* **1968**, *15*, 193–199. [CrossRef]

28. Hellenbrandt, M. The inorganic crystal structure database (ICSD)—Present and future. *Crystallogr. Rev.* **2004**, *10*, 17–22. [CrossRef]

29. Perdew, J.P.; Burke, K.; Ernzerhof, M. Generalized gradient approximation made simple. *Phys. Rev. Lett.* **1996**, *77*, 3865. [CrossRef]

30. Monkhorst, H.J.; Pack, J.D. Special points for Brillouin-zone integrations. *Phys. Rev. B* **1976**, *13*, 5188. [CrossRef]

31. Babar, S.; Weaver, J. Optical constants of Cu, Ag, and Au revisited. *Appl. Opt.* **2015**, *54*, 477–481. [CrossRef]

32. Hagemann, H.-J.; Gudat, W.; Kunz, C. Optical constants from the far infrared to the x-ray region: Mg, Al, Cu, Ag, Au, Bi, C, and Al_2O_3. *JOSA* **1975**, *65*, 742–744. [CrossRef]

33. Johnson, P.B.; Christy, R.-W. Optical constants of the noble metals. *Phys. Rev. B* **1972**, *6*, 4370. [CrossRef]

34. McPeak, K.M.; Jayanti, S.V.; Kress, S.J.; Meyer, S.; Iotti, S.; Rossinelli, A.; Norris, D.J. Plasmonic films can easily be better: Rules and recipes. *ACS Photonics* **2015**, *2*, 326–333. [CrossRef] [PubMed]

35. Olmon, R.L.; Slovick, B.; Johnson, T.W.; Shelton, D.; Oh, S.-H.; Boreman, G.D.; Raschke, M.B. Optical dielectric function of gold. *Phys. Rev. B* **2012**, *86*, 235147. [CrossRef]

36. Philipp, H.; Palik, E.D. *Handbook of Optical Constants of Solids*; Palik, E.D., Ed.; Academic: Orlando, FL, USA, 1985; Volume 749, p. 74.

37. Rioux, D.; Vallieres, S.; Besner, S.; Muñoz, P.; Mazur, E.; Meunier, M. An analytic model for the dielectric function of Au, Ag, and their alloys. *Adv. Opt. Mater.* **2014**, *2*, 176–182. [CrossRef]

38. Werner, W.S.; Glantschnig, K.; Ambrosch-Draxl, C. Optical constants and inelastic electron-scattering data for 17 elemental metals. *J. Phys. Chem. Ref. Data* **2009**, *38*, 1013–1092. [CrossRef]

39. Windt, D.L.; Cash, W.C.; Scott, M.; Arendt, P.; Newnam, B.; Fisher, R.; Swartzlander, A. Optical constants for thin films of Ti, Zr, Nb, Mo, Ru, Rh, Pd, Ag, Hf, Ta, W, Re, Ir, Os, Pt, and Au from 24 Å to 1216 Å. *Appl. Opt.* **1988**, *27*, 246–278. [CrossRef]

40. Blaber, M.; Arnold, M.; Ford, M. Designing materials for plasmonic systems: The alkali–noble intermetallics. *J. Phys. Condens. Matter* **2010**, *22*, 095501. [CrossRef]

41. Ordal, M.A.; Bell, R.J.; Alexander, R.W.; Newquist, L.A.; Querry, M.R. Optical properties of Al, Fe, Ti, Ta, W, and Mo at submillimeter wavelengths. *Appl. Opt.* **1988**, *27*, 1203–1209. [CrossRef]

42. Zeman, E.J.; Schatz, G.C. An accurate electromagnetic theory study of surface enhancement factors for silver, gold, copper, lithium, sodium, aluminum, gallium, indium, zinc, and cadmium. *J. Phys. Chem.* **1987**, *91*, 634–643. [CrossRef]

43. Jain, C.; Tuniz, A.; Reuther, K.; Wieduwilt, T.; Rettenmayr, M.; Schmidt, M.A. Micron-sized gold-nickel alloy wire integrated silica optical fibers. *Opt. Mater. Express* **2016**, *6*, 1790–1799. [CrossRef]

44. McPherson, D.J.; Supansomboon, S.; Zwan, B.; Keast, V.J.; Cortie, D.L.; Gentle, A.; Dowd, A.; Cortie, M.B. Strategies to control the spectral properties of Au–Ni thin films. *Thin Solid Films* **2014**, *551*, 200–204. [CrossRef]

45. Lonergan, W.W.; Vlachos, D.G.; Chen, J.G. Correlating extent of Pt–Ni bond formation with low-temperature hydrogenation of benzene and 1, 3-butadiene over supported Pt/Ni bimetallic catalysts. *J. Catal.* **2010**, *271*, 239–250. [CrossRef]

46. Liu, Z.; Hu, J.E.; Wang, Q.; Gaskell, K.; Frenkel, A.I.; Jackson, G.S.; Eichhorn, B. PtMo Alloy and MoO x@ Pt Core–Shell Nanoparticles as Highly CO-Tolerant Electrocatalysts. *J. Am. Chem. Soc.* **2009**, *131*, 6924–6925. [CrossRef] [PubMed]

47. Ehteshami, S.M.M.; Jia, Q.; Halder, A.; Chan, S.; Mukerjee, S. The role of electronic properties of Pt and Pt alloys for enhanced reformate electro-oxidation in polymer electrolyte membrane fuel cells. *Electrochim. Acta* **2013**, *107*, 155–163. [CrossRef]

48. Takahashi, H.; Niidome, Y.; Niidome, T.; Kaneko, K.; Kawasaki, H.; Yamada, S. Modification of gold nanorods using phosphatidylcholine to reduce cytotoxicity. *Langmuir* **2006**, *22*, 2–5. [CrossRef] [PubMed]

49. Jain, P.K.; Lee, K.S.; El-Sayed, I.H.; El-Sayed, M.A. Calculated absorption and scattering properties of gold nanoparticles of different size, shape, and composition: Applications in biological imaging and biomedicine. *J. Phys. Chem. B* **2006**, *110*, 7238–7248. [CrossRef]

50. Gao, J.; Bender, C.M.; Murphy, C.J. Dependence of the gold nanorod aspect ratio on the nature of the directing surfactant in aqueous solution. *Langmuir* **2003**, *19*, 9065–9070. [CrossRef]

51. Link, S.; Mohamed, M.; El-Sayed, M. Simulation of the optical absorption spectra of gold nanorods as a function of their aspect ratio and the effect of the medium dielectric constant. *J. Phys. Chem. B* **1999**, *103*, 3073–3077. [CrossRef]

 *micromachines*

Article

Voltage-Tunable Mid- and Long-Wavelength Dual-Band Infrared Photodetector Based on Hybrid Self-Assembled and Sub-Monolayer Quantum Dots

Yao Zhai [1,†], Guiru Gu [2] and Xuejun Lu [1,*]

[1] Department of Electrical and Computer Engineering, University of Massachusetts Lowell, One University Avenue, Lowell, MA 01854, USA; Yao.Zhai@colorado.edu

[2] Department of Physics, Stonehill College, 320 Washington Street, Easton, MA 02357, USA; ggu@stonehill.edu

* Correspondence: Xuejun_Lu@uml.edu

† Current Address: Department of Mechanical Engineering, University of Colorado Boulder, Boulder, CO 80309, USA.

Received: 27 November 2018; Accepted: 17 December 2018; Published: 22 December 2018

Abstract: In this paper, we report a mid-wave infrared (MWIR) and long-wave infrared (LWIR) dual-band photodetector capable of voltage-controllable detection band selection. The voltage-tunable dual-band photodetector is based on the multiple stacks of sub-monolayer (SML) quantum dots (QDs) and self-assembled QDs. By changing the photodetector bias voltages, one can set the detection band to be MWIR, or LWIR or both with high photodetectivity and low crosstalk between the bands.

Keywords: submonolayer quantum dots; self-assembled quantum dot; dual-band midwave and longwave infrared photodetector; voltage tunable detection spectrum

1. Introduction

Due to their importance in numerous civilian and defense applications, dual-band photodetectors covering the mid-wave infrared (MWIR) (3–5 μm) and long-wave infrared (LWIR) (8–12 μm) bands have been extensively researched in the past decades [1,2]. The current state-of-the-art infrared detection technology is based on the HgCdTe alloy. By modifying Hg to Cd component ratio in the alloy [3], one can tune the detection spectrum from 1 to 25 μm. However, due to the difficulties of growing large size HgCdTe wafers with high quality and good uniformity, HgCdTe based IR photodetectors and focal plane arrays (FPAs) are not only very expensive and but also very hard to make, especially for large format (1024 × 1024) FPAs. Graphene photodetectors are a new emerging technology that could offer broadband detection, high responsivity, and high operating temperatures [4]. One of the limitations, however, is the availability of large area graphene for large format (1024 × 1024) FPA development.

Quantum-dot infrared photodetector (QDIP) technology has been extensively studied as a promising large area IR sensing and imaging technology due to the advantages provided by the Quantum-dot (QD) nanostructures, including the three-dimensional (3D) quantum confinement that modifies the quantum selection rules to allow surface normal incident light detection [5], low dark current [6], high photoresponsivity [7], and high operating temperature [7–9]. QD based MWIR and LWIR dual-band photodetectors have been reported by several research groups [10–12] using different techniques including tailoring the sizes of the QDs during the material growth capping the QDs with different capping layers [12], and inserting asymmetric barrier layers [13]. Most of these dual band QDIPs are based on self-assembled QDs grew via the Stranski-Krastanow (S-K) growth mode (referred to as the self-assembled QDs henceforth). Submonolayer (SML) QD growth is an alternative QD growth mode [14–16], where only a fraction of a monolayer (ML) of InAs is grown followed by the growth of the GaAs separation layer. QDs form by growing multiple repeats of the SML InAs with the GaAs separation layer (referred to as the SML QDs henceforth) [14–16]. SML based QDs offer a few advantages, including smaller base width and absence of wetting layer, large density of QDs, reduced strain for more QD layers, and engineerable correlation between the vertically stacked QD layers by varying the thickness of the separation layers [15,17,18]. In addition, since the thicknesses of the SML InAs layer and its GaAs separation layer can be adjusted individually, the SML QD technique provides additional flexibility in designing the detection spectrum for multispectral sensing and imaging. Due to these advantages, SML QD based optoelectronic devices have been extensively researched, including lasers [16] and infrared (IR) photodetectors [11,17].

In this paper, we report a dual-band MWIR and LWIR photodetector based on the multiple stacks of SML QDs and self-assembled QDs. By engineering the SML and the self-assembled QDs, we obtained MWIR and LWIR dual-band detection with low crosstalk between the MWIR and the LWIR bands. The MWIR and LWIR dual-band photodetector also shows voltage-controllable band selection. By changing the photodetector bias voltages, one can set the detection band to be MWIR, or LWIR or both with low crosstalk and high photodetectivity.

2. Device Structure

Figure 1a shows the schematic structure of the hybrid QDIP. It consists of 10 layers of SML QDs and 10 layers of self-assembled QDs sandwiched between the top and bottom contact layers. The total thickness of the SML and the self-assembled QD layers is 1.5 μm. Figure 1b shows the simplified conduction band structure of the hybrid SML and the self-assembled QDIP. The QDIP is a photoconductor where the MWIR/LWIR incident light excites the electrons from the ground states to the excited states. The electrons are subsequently collected by the top and bottom electrodes and form photocurrent. Due to different strains in forming the SML QDs and self-assembled QDs, their sizes and the energy levels are unalike and thus allow them to detect different bands. By optimizing the QD growth conditions such as the growth rate, substrate temperatures, and the V to III ratios, one can achieve the detection of the MWIR and the LWIR bands using the self-assembled and SML QD groups, respectively. The growth strain-driven band structure engineering offers an alternative way to achieve dual-band photodetectors [19].

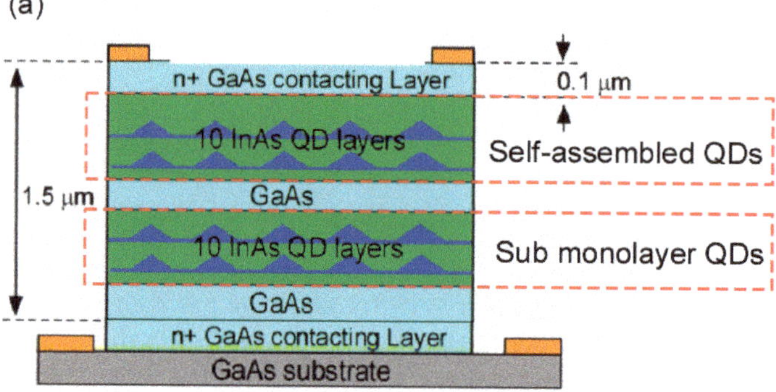

Figure 1. *Cont.*

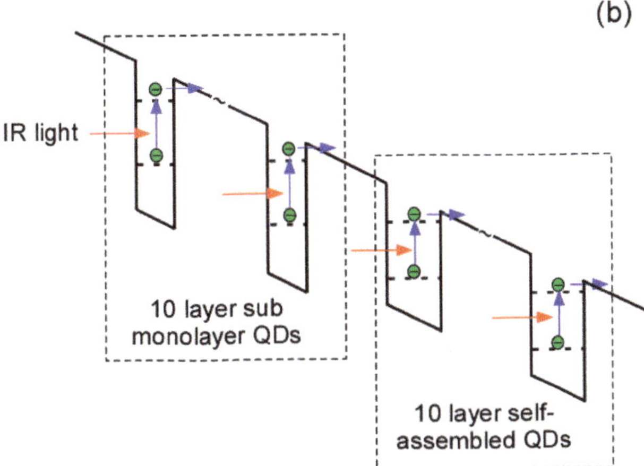

Figure 1. (**a**) Schematic structures of the hybrid Quantum-dot infrared photodetector (QDIP). It consists of the Submonolayer (SML) QDs and self-assembled QDs sandwiched between the top and the bottom contacts. (**b**) Simplified conduction band structure of the SML and self-assembled QDIP. Due to the different energy levels, the SML and self-assembled QDs can detect different bands.

3. Material Growth and Device Fabrication Process

The multiple stacks of SML QD and self-assembled QD based dual band photodetector structure was grown on a semi-insulating GaAs (100) substrate using a V80H molecular beam epitaxy (MBE) system. A 3000 Å undoped GaAs buffer layer was first grown on the substrate, followed by the growth of a 1500 Å n+ doped GaAs bottom contact layer. Another 1000 Å undoped GaAs buffer layer was then grown on the bottom contact layer, followed by the growth of the 10 period SML QD layers. Each period of the SML QD layer consists of 10 repeats of the InAs 0.45 mL and 1 mL GaAs stacks. A 20 Å of $Al_{0.1}Ga_{0.9}As$ and a 450 Å GaAs spacer layer were grown between the two adjacent SML QD layers as the separation layer.

Figure 2 shows an atomic force microscopy (AFM) (Park Systems, Suwon, Korea) image of SML QDs grown under the same growth conditions as the photodetector device. The SML QDs are quite uniform with a high density of $\sim 4 \times 10^{10}$/layer.

After the SML QD growth, 10 periods of the self-assembled QD layers were grown on the SML QD layers. Each period of the self-assembled QD heterostructures consists of 1 nm $In_{0.15}Ga_{0.85}As$, 2 mL of InAs QDs, 6 nm $In_{0.15}Ga_{0.85}As$ cap layer and 45 nm GaAs spacer layer. A 500 Å undoped GaAs buffer and a 0.1 µm n+ doped GaAs top contact layer were finally grown to finish the MBE growth. The doping level of the SML and the self-assembled QD regions were tuned to be approximately 2 electrons per dot. The SML and the self-assembled QD layers were grown at 485 °C. All the other buffer and contact layers were grown at 580 °C.

After the growth, the wafer is then processed into 250 µm-diameter circular mesas with top and bottom electrodes using standard photolithography, wet etching procedures, E-beam metal evaporation deposition, lift-off and thermal annealing processes [7]. The finished device was mounted on a copper plate and put into a liquid nitrogen (LN_2) cooled dewar with a Zinc Selenide (ZnSe) IR window.

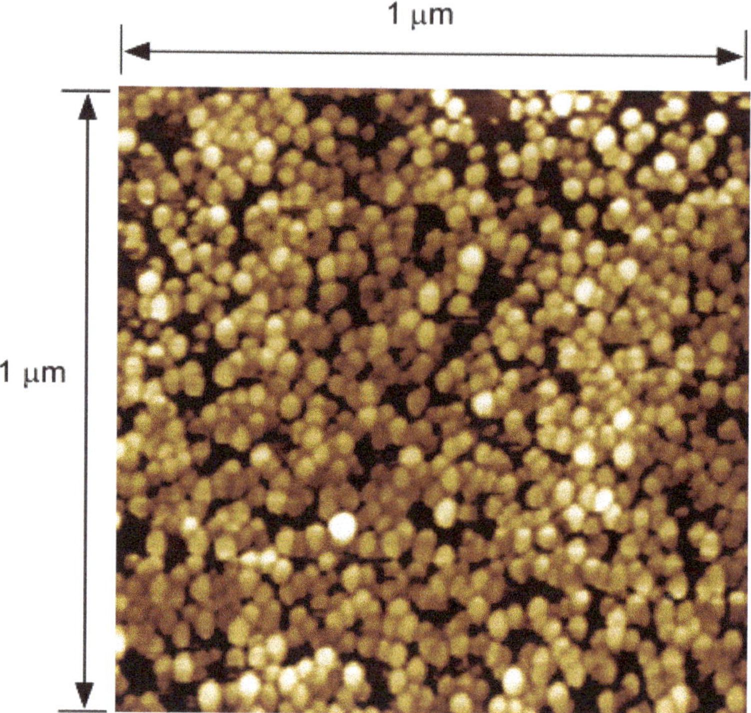

Figure 2. Atomic force microscopy (AFM) image of the QDs grown at the same growth conditions as the photodetector device. The SML QDs are uniform with a high density of ~ 4 × 10^{10}/layer.

4. Device Measurement and Characterization

The spectral response of the dual-band QDIP was measured using a Fourier transform infrared spectrometer (FTIR) (Bruker, Leipzig, Germany) at 77 K. Detailed device characterization procedures have been reported [12]. Figure 3 shows the photocurrent spectra of the dual-band QDIP under bias voltages of −3.4 V (blue trace) and +1.6 V (pink trace). At the negative bias of −3.4 V, the QDIP only shows the IR photodetection from 6.8 μm to 10.5 μm with the peak wavelength of 8.2 μm. At +1.6 V, the photodetector covers the 4.0 μm to 6.8 μm wavelength regime with the peak wavelength of 5.5 μm. The photocurrent in the 6.8 μm to 10.5 μm spectral range is much larger than that in the 4.0 μm to 6.8 μm wavelength regime.

Figure 4 shows the photocurrent spectrum of the dual-band QDIP under the bias voltage of −1.1 V. Under this bias voltage, the QDIP covers both the MWIR and the LWIR band.

The dark current (I_d) of the dual-band QDIP was measured using a Keithley source meter. Figure 5 shows the dark current density J_d at different biases measured at the device temperature of 77 K. The noise current (i_{noise}, in units of A/HzP$^{1/2P}$) was characterized using a low-noise current preamplifier (Stanford Research Systems, Sunnyvale, CA, USA) and a fast Fourier transform (FFT) spectrum analyzer (Stanford Research Systems, Sunnyvale, CA, USA). To avoid 1/f noise contributions, the noise current (i_{noise}) was determined at 597 Hz.

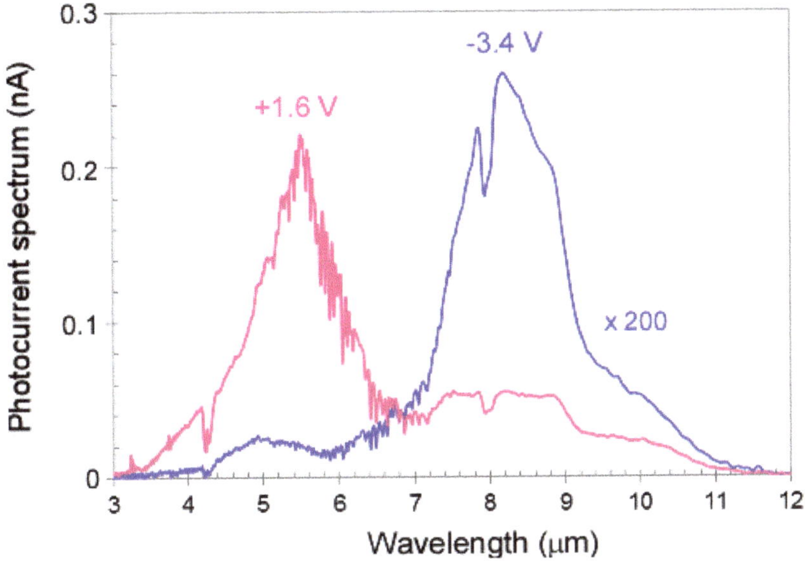

Figure 3. Photocurrent spectra of the dual-band QDIP under bias voltages of −3.4 V (blue trace) and +1.6 V (pink trace). The negative bias gives the LWIR band, whereas under the positive bias, the MWIR band dominates.

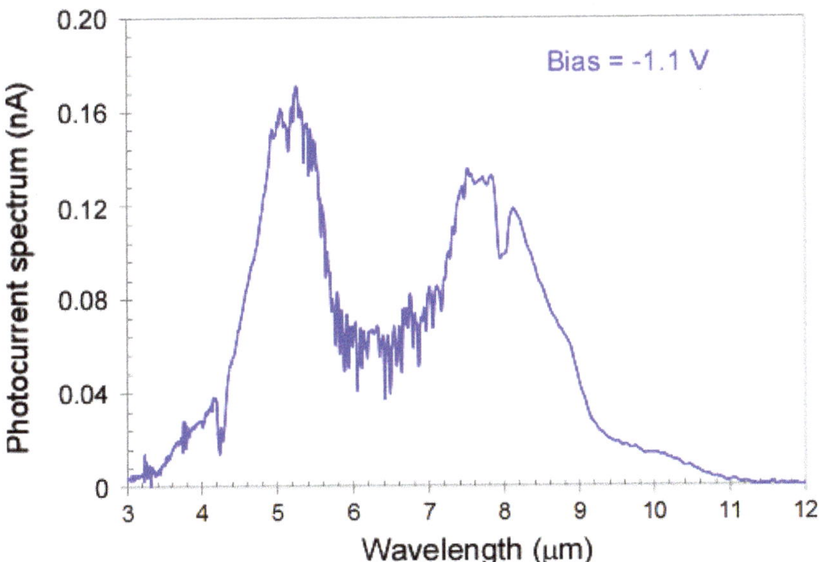

Figure 4. Photocurrent spectrum of the dual-band QDIP under bias voltages of −1.1 V. At this bias, the QDIP covers both the MWIR and the LWIR bands.

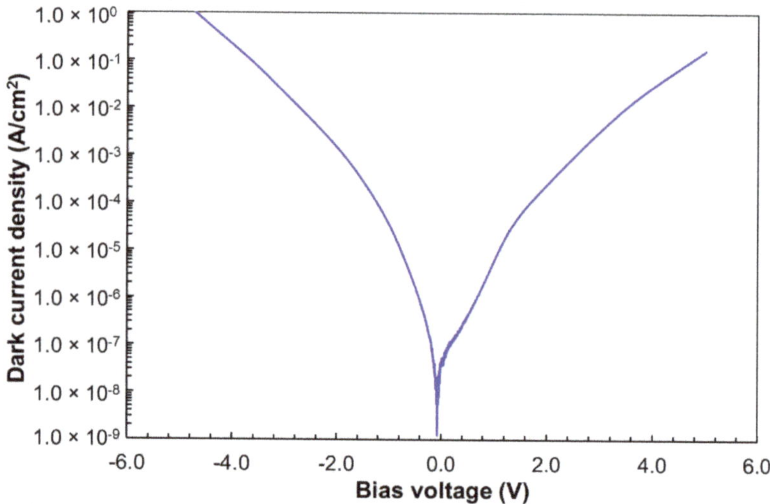

Figure 5. Dark current density (J_d) of the dual-band QDIP under different bias voltages.

The photoresponsivity $\Re$ for the MWIR and LWIR bands were measured using a (1000 K) calibrated blackbody source, with 3–5 µm and 8–12 µm band-pass filters. Since the detector bands are not exactly aligned with the band-pass filters, this would cut some IR signals and thus under-estimate the photoresponsivity of both bands. Nevertheless, we still observed the dual-band detection and voltage-controllable band tuning with good selectivity. The photocurrents were measured by modulating the filtered blackbody infrared source at a chopper frequency of 597 Hz, and the signal current was collected using a low-noise current preamplifier and an FFT spectrum analyzer.

Figure 6 shows the photoresponsivity $\Re$ of the dual-band QDIP under different bias voltages.

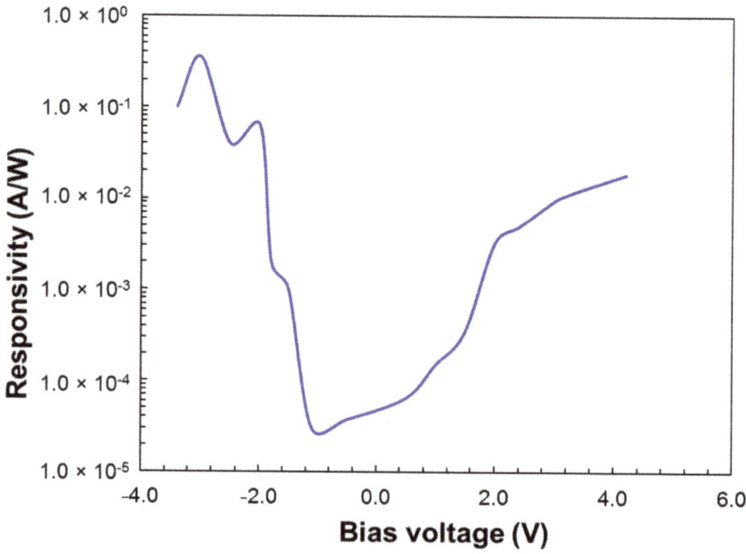

Figure 6. Photoresponsivity $\Re$ of the dual-band QDIP under different bias voltages.

The photodetectivity D^* can be calculated using:

$$D^* = \frac{\Re\sqrt{A}}{\sqrt{i^2_{noise}}}$$

(1)

where, A is the detector area, $\Re$ is the photoresponsivity, and i_{noise} is the noise current.

Figure 7 shows the calculated photodetectivity (D^*) at different bias voltages with the 3–5 μm (square) and 8–12 μm (triangle) band pass filters. The 8–12 μm band has a higher photodetectivity (D^*) at negative biases, whereas the photodetectivity D^* of the 3–5 μm band is higher at positive biases. The photodetectivity D^* differences indicate good band selectivity by direct bias voltage control.

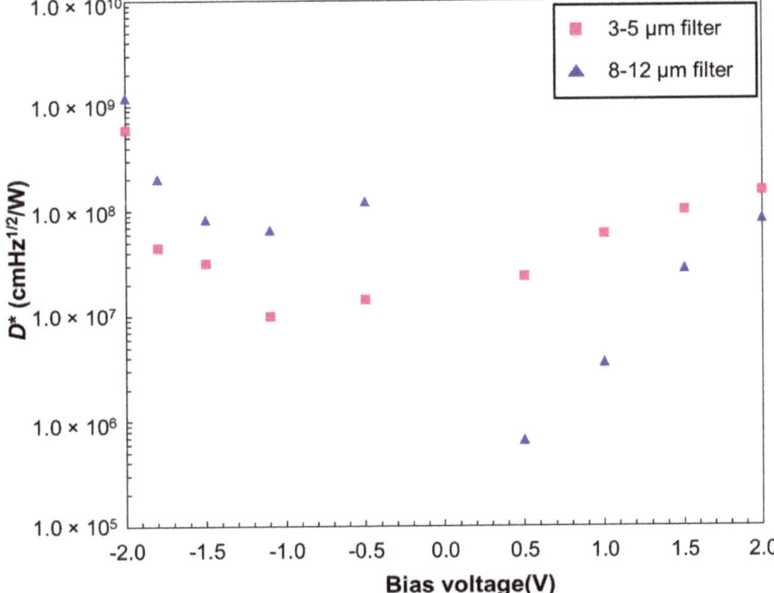

Figure 7. Calculated photodetectivity (D^*) at different bias voltages with the 3–5 μm (square) and 8–12 μm (triangle) band pass filters. The 8–12 μm band has a higher D^* at negative biases, whereas the photodetectivity D^* of the 3–5 μm band is higher at positive biases.

5. Conclusions

In conclusion, we report a dual-band QDIP based on vertically stacked SML and self-assembled QDs. The dual-band QDIP covers both the MWIR and LWIR spectral regimes. The dual-band QDIP also shows voltage-tunable detection band selection capability. By changing the photodetector bias voltages, one can set the detection band to be MWIR, or LWIR or both with high photodetectivity and low crosstalk between the bands.

Author Contributions: X.L. conceived and designed the experiments; Y.Z. and G.G. performed the experiments; Y.Z., G.G. and X.L. analyzed the data; All contributed to the paper writing.

Acknowledgments: This research was partially supported by the Air Force Office of Scientific Research (AFOSR) under contract No. FA9550-12-1-0176. The authors appreciate Applied NanoFemto Technologies LLC (ANFT)'s support of MBE growth of the SML and the self-assembled QD samples. The devices were fabricated using University of Massachusetts Lowell's Core Research Facilities (CRF).

Conflicts of Interest: Xuejun Lu is a co-founder of ANFT. The authors declare no conflict of interest.

References

1. Ballingall, R.A. Review of infrared focal plane arrays. *Infrared Technol. Appl.* **1990**, *1320*, 70–87.
2. Levine, B.F. Quantum-well infrared photodetectors. *J. Appl. Phys.* **1993**, *74*, 1–81. [CrossRef]
3. Norton, P. HgCdTe infrared detectors. *Opto-Electron. Rev.* **2002**, *10*, 159–174.
4. Wang, J.; Cheng, Z.; Chen, Z.; Xu, J.-B.; Tsang, H.K.; Shu, C. Graphene photodetector integrated on silicon nitride waveguide. *J. Appl. Phys.* **2015**, *117*, 144504. [CrossRef]
5. Pan, D.; Towe, E.; Kennerly, S. Normal-incidence intersubband (In, Ga)As/GaAs quantum dot infrared photodetectors. *Appl. Phys. Lett.* **1998**, *73*, 1937–1939. [CrossRef]
6. Vaillancourt, J.; Stintz, A.; Meisner, M.J.; Lu, X. Low-bias, high-temperature operation of an InAs–InGaAs quantum-dot infrared photodetector with peak-detection wavelength of 11.7 μm. *Infrared Phys. Technol.* **2009**, *52*, 22–24. [CrossRef]
7. Lu, X.; Vaillancourt, J.; Meisner, M.J. Temperature-dependent photoresponsivity and high-temperature (190 K) operation of a quantum dot infrared photodetector. *Appl. Phys. Lett.* **2007**, *91*, 051115. [CrossRef]
8. Stiff, A.D.; Krishna, S.; Bhattacharya, P.; Kennerly, S. High-detectivity, normal-incidence, mid-infrared (λ~4 μm)InAs/GaAs quantum-dot detector operating at 150 K. *Appl. Phys. Lett.* **2001**, *79*, 421–423. [CrossRef]
9. Vaillancourt, J.; Vasinajindakaw, P.; Hong, W.; Lu, X.; Qian, X.; Vangala, S.R.; Goodhue, W.D. A LWIR quantum dot infrared photodetector working at 298K. In Proceedings of the SPIE OPTO, San Francisco, CA, USA, 21–25 February 2010; pp. 760821–760826.
10. Ye, Z.; Campbell, J.C.; Chen, Z.; Kim, E.-T.; Madhukar, A. Voltage-controllable multiwavelength InAs quantum-dot infrared photodetectors for mid- and far-infrared detection. *J. Appl. Phys.* **2002**, *92*, 4141–4143. [CrossRef]
11. Kim, J.O.; Sengupta, S.; Sharma, Y.; Barve, A.V.; Lee, S.J.; Noh, S.K.; Krishna, S. Sub-monolayer InAs/InGaAs quantum dot infrared photodetectors (SML-QDIP). In Proceedings of the SPIE Defense, Security, and Sensing, Baltimore, MD, USA, 23–27 April 2012; p. 835336.
12. Meisner, M.J.; Vaillancourt, J.; Lu, X. Voltage-tunable dual-band InAs quantum-dot infrared photodetectors based on InAs quantum dots with different capping layers. *Semicond. Sci. Technol.* **2008**, *23*, 095016. [CrossRef]
13. Huang, J.; Ma, W.; Wei, Y.; Zhang, Y.; Huo, Y.; Cui, K.; Chen, L. Two-color In 0.4 Ga 0.6 As/Al 0.1 Ga 0.9 As quantum dot infrared photodetector with double tunneling barriers. *Appl. Phys. Lett.* **2011**, *98*, 103501. [CrossRef]
14. Krestnikov, I.L.; Ledentsov, N.N.; Hoffmann, A.; Bimberg, D. Arrays of Two-Dimensional Islands Formed by Submonolayer Insertions: Growth, Properties, Devices. *Phys. Status Solidi A* **2001**, *183*, 207–233. [CrossRef]
15. Xu, Z.; Birkedal, D.; Hvam, J.M.; Zhao, Z.; Liu, Y.; Yang, K.; Kanjilal, A.; Sadowski, J. Structure and optical anisotropy of vertically correlated submonolayer InAs/GaAs quantum dots. *Appl. Phys. Lett.* **2003**, *82*, 3859–3861. [CrossRef]
16. Hopfer, F.; Mutig, A.; Kuntz, M.; Fiol, G.; Bimberg, D.; Ledentsov, N.N.; Shchukin, V.A.; Mikhrin, S.S.; Livshits, D.L.; Krestnikov, I.L.; et al. Single-mode submonolayer quantum-dot vertical-cavity surface-emitting lasers with high modulation bandwidth. *Appl. Phys. Lett.* **2006**, *89*, 141106. [CrossRef]
17. Ting, D.Z.-Y.; Bandara, S.V.; Gunapala, S.D.; Mumolo, J.M.; Keo, S.A.; Hill, C.J.; Liu, J.K.; Blazejewski, E.R.; Rafol, S.B.; Chang, Y.-C. Submonolayer quantum dot infrared photodetector. *Appl. Phys. Lett.* **2009**, *94*, 111107. [CrossRef]
18. Ning, L.; Peng, J.; Zhan-Guo, W. Broadband light emitting from multilayer-stacked InAs/GaAs quantum dots. *Chin. Phys. B* **2012**, *21*, 117305.
19. Perera, A.; Ariyawansa, G.; Huang, G.; Bhattacharya, P. Quantum dot nanostructures for multi-band infrared detection. *Infrared Phys. Technol.* **2009**, *52*, 252–256. [CrossRef]

 *micromachines*

Article

A Trace Carbon Monoxide Sensor Based on Differential Absorption Spectroscopy Using Mid-Infrared Quantum Cascade Laser

Chen Chen, Qiang Ren, Heng Piao, Peng Wang and Yanzhang Wang *

College of Instrumentation & Electrical Engineering, Key Laboratory of Geophysical Exploration Equipment, Ministry of Education of China, Jilin University, Changchun 130026, China; cchen@jlu.edu.cn (C.C.); renqiang15@mails.jlu.edu.cn (Q.R.); piaoheng18@mails.jlu.edu.cn (H.P.); wangpeng18@mails.jlu.edu.cn (P.W.)
* Correspondence: yanzhang@jlu.edu.cn

Received: 25 November 2018; Accepted: 14 December 2018; Published: 18 December 2018

Abstract: Carbon monoxide (CO), as a dangerous emission gas, is easy to accumulate in the complex underground environment and poses a serious threat to the safety of miners. In this paper, a sensor using a quantum cascade laser with an excitation wavelength of 4.65 µm as the light source, and a compact multiple reflection cell with a light path length of 12 m is introduced to detect trace CO gas. The sensor adopts the long optical path differential absorption spectroscopy technique (LOP-DAST) and obtains minimum detection limit (MDL) of 108 ppbv by comparing the residual difference between the measured spectrum and the Voigt theoretical spectrum. As a comparison, the MDL of the proposed sensor was also estimated by Allan deviation; the minimum value of 61 ppbv is achieved while integration time is 40 s. The stability of the sensor can reach 2.1×10^{-3} during the 2 h experimental test and stability of 1.7×10^{-2} can still be achieved in a longer 12 h experimental test.

Keywords: Trace carbon monoxide sensor; mid-infrared spectrum; quantum cascade laser; differential absorption spectroscopy; residual analysis

---

1. Introduction

Compared to traditional bipolar semiconductor lasers, quantum cascade lasers (QCLs) offer unique advantages of good monochromaticity, high quantum efficiency, good temperature stability, flexible wavelength design, and fast response [1–3]. In addition, the infrared spectrum of QCL covers three important atmospheric transmission windows. Therefore, QCLs have an incomparable advantage over other luminous sources in the field of gas detection [4–6].

In recent years, the application of QCL to detect trace gas in the infrared "fingerprint region" has been developing rapidly. In 2008, J. B. Mcmanus et al. at the Aerodyne research center, adopted a QCL with center wave number of 967 cm^{-1} combined with a long optical path direct absorption method to detect ammonia gas. The light path length of multiple reflecting cell was increased to 76 m, and the minimum detection limit (MDL) of ammonia gas was achieved as 0.2 ppbv [7]. In 2013, P. G. Carbajo et al. detected H_2CO using a QCL with central wave number of 1769 cm^{-1}. The MDL was improved to 0.06 ppbv by using the optical feedback-cavity enhanced absorption spectrum technology [8]. In 2010, K. Ruifeng et al. utilized the pulsed QCL with the central wave number of 1904 cm^{-1} combined with direct absorption spectrum detection technology to detect nitric oxide gas [9]. The MDL of nitric oxide reached 3.4 ppmv by performing the least squares calculation method to fit baseline. In 2014, L. Guolin et al. reported a CO_2 sensor using the 4.8 µm mid-infrared QCL to realize a MDL of 180 ppbv [10].

The optimization work on open gas cell using ellipsoid condensers is the main contribution here. Although the mentioned sensors achieved good detection performance for target gases, the shortcoming of complicated structure limits their usage in field applications. In this paper, a new CO sensor using mid-infrared QCL with center wavelength of 4.65 μm is introduced. The sensor utilizes a 12 m long optical path differential absorption spectroscopy technique (LOP-DAST) to achieve MDL of 108 ppbv, minimum Allan deviation of 61 ppbv, and high performance of working stability during the long experimental test. To resolve the issue of large packages of core constituted components, chip-level mid-infrared gas sensors with high sensitivity is discussed in the final section, which appears very promising for future gas sensing.

2. Detection Principle of CO Using LOP-DAST

2.1. Selection of CO Absorption Line

In the mid-infrared spectrum band, the absorption spectrum of gas molecules generally includes rotational spectrum and rotational vibration spectrum, and each gas has many absorption bands in the mid-infrared spectrum band [11–13]. Taking CO as a target, the absorption spectrum band around 4.6 μm caused by the fundamental frequency vibration of CO molecules is shown in Figure 1.

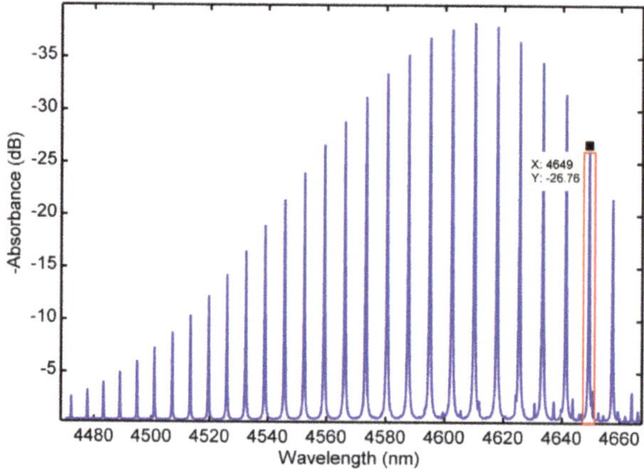

Figure 1. Absorption spectrum of CO.

As shown in Figure 1, the x coordinate is wavelength (nanometers, nm), the y coordinate is absorption intensity (decibel, dB), and blue lines represent CO gas absorption lines. Due to the limitation of tuning ability of mid-infrared QCL adopted in this paper, the optimum luminous wavelength is chosen at 4.65 μm, which is close to the peak value of CO absorption spectrum at 4.649 μm, as illustrated in the red box. For this absorption spectrum line, the absorption coefficient of CO is −26 dB and is in the atmospheric window range. In addition, the improved accuracy of measured results can be reached because of spectral interference exclusion from other atmospheric components, such as water, methane, and CO_2.

2.2. Derivation of LOP-DAST technique

The LOP-DAST can be described as follows. According to the Beer-Lambert law, absorbance is proportional to the concentrations of the attenuating species in the material sample [14–18]. The light intensity after passing through the gas absorption cell with light path length L is:

$$I(\lambda) = I_0(\lambda)exp(\phi(\sigma_i, C_i, L, \varepsilon_r, \varepsilon_m)) \tag{1}$$

The multiple variables function in the above equation can be expressed as:

$$\phi(\sigma_i, C_i, L, \varepsilon_r, \varepsilon_m) = \sum_{i=1}^{n} \sigma_i(\lambda)C_iL + \varepsilon_r(\lambda) + \varepsilon_m(\lambda) \tag{2}$$

Here, λ is the wavelength of luminous light, $\sigma_i(\lambda)$ denotes the absorption cross section of the ith gas, C_i is the average concentration of the ith gas, L is the light path length, $\varepsilon_r(\lambda)$ is Rayleigh scattering, $\varepsilon_m(\lambda)$ is Mie scattering. The absorption cross section of Rayleigh scattering and Mie scattering are shown as a broadband absorption cross section. The LOP-DAST method decomposes the absorption cross section into broadband absorption cross section and narrow absorption cross section, namely:

$$\sigma_i(\lambda) = \sigma_i^b(\lambda) + \sigma_i^l(\lambda) \tag{3}$$

where, $\sigma_i^b(\lambda)$ represents the broadband absorption cross section of the ith gas, and $\sigma_i^l(\lambda)$ represents the narrow absorption cross section of the ith gas. Therefore, Formula (2) can be divided into the following form:

$$\phi^b\left(\sigma_i^b, C_i, L, \varepsilon_r, \varepsilon_m\right) = \sum_{i=1}^{n} \sigma_i^b(\lambda)C_iL + \varepsilon_r(\lambda) + \varepsilon_m(\lambda) \tag{4}$$

$$\phi'\left(\sigma_i^l, C_i, L\right) = \sum_{i=1}^{n} \sigma_i^l(\lambda)C_iL \tag{5}$$

By means of numerical filtering, the terms ϕ^b only containing broadband absorption cross section can be removed, so as to obtain the differential optical density of the measured gas,

$$(O.D.)_\lambda = \ln\left[\frac{I(\lambda)}{I_0(\lambda)}\right] = \sum \sigma_i^l(\lambda)C_iL \tag{6}$$

The CO concentration in absorption cell can be obtained by fitting the differential optical density with the reference spectrum via the least square method.

The above analysis shows that obtaining the reference absorption cross section in the absorption spectrum band is key to measuring the CO concentration. The CO absorption line shape at ambient temperature and a bar pressure given by HITRAN database are used to calculate the Voigt theoretical spectrum. The Voigt theoretical spectrum takes into account the Lorentz lineshape generated by gas collision, spontaneous radiation and Doppler lineshape generated by velocity distribution of luminous particles. The comprehensive function is as follows:

$$\begin{aligned} g_z(v, v_0) &= \int_{-\infty}^{+\infty} g_L(v_1, v_0)g_D(v, v_1)dv_1 \\ &= \frac{1}{\pi}\sqrt{\frac{\ln 2}{\pi}} \int_{-\infty}^{+\infty} \frac{\alpha_L}{(v_1-v_0)^2 - \alpha_L^2} \frac{1}{\alpha_D}exp(-\omega(v_1))dv_1 \end{aligned} \tag{7}$$

where,

$$\omega(v_1) = \frac{\ln 2}{\alpha_D^2}(v - v_1)^2 \tag{8}$$

α_L and α_D are the half-width of Lorentz lineshape and Doppler lineshape, respectively. v_0 is the central frequency. If α_D is set as a constant [19], then:

$$g_z(v, v_0) = \sqrt{\frac{\ln 2}{\pi}}\frac{1}{\alpha_D}\left(\frac{\mu}{\pi}\int_{-\infty}^{+\infty}\frac{1}{(\xi - t)^2 - \mu^2}exp\left(-t^2\right)dt\right) \tag{9}$$

where,

$$\xi = \sqrt{\ln 2}\frac{v - v_0}{\alpha_D}, \quad \mu = \sqrt{\ln 2}\frac{\alpha_L}{\alpha_D}, \quad t = \sqrt{\frac{\ln 2}{\alpha_D}}(v_1 - v) \tag{10}$$

The Voigt absorption spectrum has been widely used in differential absorption spectrometry. The MDL of the sensor can be determined by comparing the residual between the spectrum measured in the experiment and the Voigt theoretical spectrum.

3. Sensor Configuration

The core device of a gas sensor is a QCL with a center wavelength of 4.65 µm. The luminous mid-infrared light is focused on the 12 m long optical path absorption cell (made of stainless steel) through the two reflected-gilt spherical lens. Then, the mid-infrared light emitted from the absorption cell passes through a paraboloid reflector and is received by a liquid nitrogen cooled HgCdTe detector; the response peak wavelength is 5 µm, the spectral response range is from 2 µm to 12 µm, the response time is less than 100 ns, and the size of the detection surface is 1 mm^2. The overall design block diagram and physical image of the CO sensor is shown in Figure 2a,b, respectively.

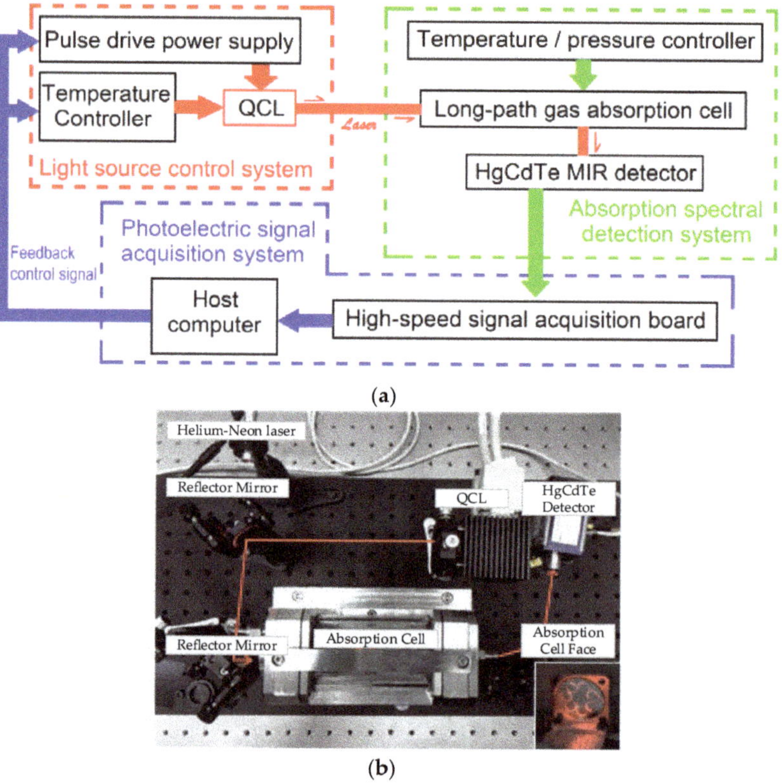

Figure 2. (a) The block diagram of CO sensor. (b) Physical image of CO sensor.

In the control system, the self-designed constant power controller for QCL is utilized, which is capable of the adjustment of width and duty ratio of the output driving pulse. The peak driving current is 10 A, and the pulse rising/falling time is less than 10 ns. The laser is real-time controlled in a constant power model by reading the feedback current from the photodiode packaged in the inner QCL. The QCL working temperature fluctuation is less than ±0.05 °C by using adaptive proportional integral derivative (PID) algorithm, and no overshoot control of target working temperature is realized.

The optical part is composed of a gold-plated reflector, multiple reflection mirror, and a mid-infrared HgCdTe detector. The reflector selects platinum materials as the coating, which have a higher than 99% reflectivity for mid-infrared light. The light path length of the absorption cell used in the sensor is 12 m, the volume is 500 mL, the operating reflection band range is 2.5 µm to 10 µm, and the physical size is 450 mm × 110 mm × 110 mm. Due to the invisibility of mid-infrared, the luminous light emitted from QCL is collimated via a visible red light from the Helium-Neon laser. As a result, most of the mid-infrared light can be reflected through an all-reflecting spherical lens into the inlet aperture of the absorption cell. Finally, optical output signals can be detected by the HgCdTe detector after several passes in the absorption cell.

4. Experiment

Under the conditions of ambient temperature (293 K) and one bar pressure (1.0 × 10^5 Pa), the performance indicators of trace CO sensor, the concentration-voltage response, measurement MDL, and working stability are tested. MDL performance is assessed by comparing the residual difference method and the Allan deviation method.

4.1. Response

In order to observe the response of the proposed sensor, 15 CO gas samples with different standard concentrations (10 ppmv to 60 ppmv) were prepared by dynamic gas dilution equipment. The prepared CO gases are pumped into the gas cell successively, and the relation between output voltage signal of the sensor and CO concentration was obtained, as shown in Figure 3.

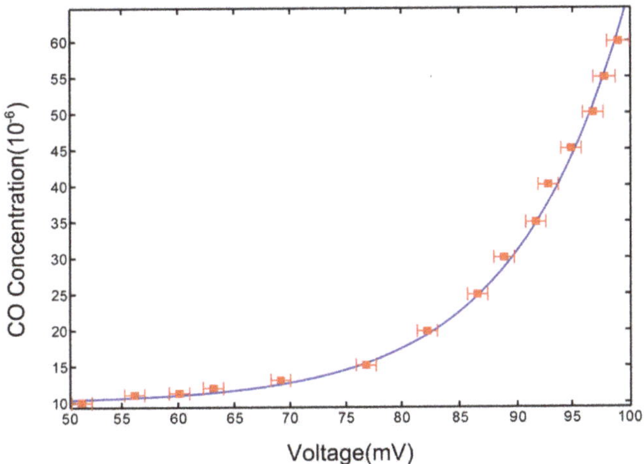

Figure 3. Concentration of CO versus output voltage of the sensor.

In the figure, the x coordinate represents the output voltage collected by the photoelectric signal acquisition system, the y coordinate describes the concentration of the CO gas sample, and the solid blue line is the fitting curve. It can be seen from the figure that the measured gas concentration with respect to the output voltage is exponential, which is in line with Bill lambert's law, as shown in Formula (1).

4.2. Stability

Stability performance can be obtained by measuring the sensor's output changes along with time and a certain concentration of measured gas in absorption cell. The CO gas with a concentration of 60 ppmv was observed for 12 h, and the relationship between the output voltage corresponding to CO gas concentration and the detection time was recorded; the results are shown in Figure 4.

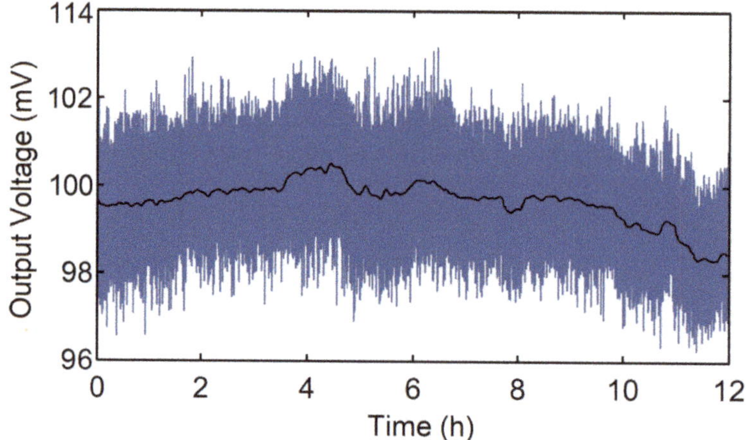

Figure 4. The measured data of 60 ppmv CO over 12 h.

The blue curve is the output voltage collected per second, and the black curve is the average value of the collected data with integral time of 30 s. It can be seen from the figure that the sensor had stable operations within two hours of starting up, the recorded data fluctuated with 0.4 mV, and stability was as high as 2.1×10^{-3}. Accordingly, the response amplitude of the sensor changed within 1.6 mv in 12 h, and the stability of 1.7×10^{-2} was achieved in the longer experimental test. In future work, reference cells can adapt to suppress common noise, which leads to stability promotion of the proposed sensor.

4.3. Minimum Detection Limit

MDL is defined as the gas concentration when the effective gas absorption signal is distinguishable from background noise, namely when Signal-Noise-Ratio (SNR) is 1. MDL of the proposed sensor was estimated by comparing the residual difference between the measured spectrum and the Voigt theoretical spectrum. The experimental results to measure 10 ppmv CO is shown in Figure 5.

The red curve in Figure 5a is the Voigt theoretical spectrum; the difference curve between the measured absorption spectrum and the base line background is shown below (blue line). The coincidence degree between the Voigt theoretical spectrum and the difference curve is very good, the residual error less than ±0.5%. We investigated the noise spectrum near the absorption peak and found that the noise standard error (SE) is 9.395×10^{-4}. Here, we can approximately assume that this SE value is the minimum detectable concentration when the Signal-Noise-Ratio (SNR) equals to 1. In addition, the magnitude of absorption spectrum peak is 0.087, so the SNR = 0.087/SE = 92.6. Due to the measured CO concentration being 10 ppmv, the MDL of the sensor can be calculated as 10 ppmv/92.6 = 108 ppbv.

Due to the characteristics of non-stationary and slow-time variation when the sensor worked, we introduced Allan deviation to estimate MDL more accurately. Allan deviation has been widely used in the estimation of MDL of gas sensor [20,21], and its expression is:

$$\sigma_A^2(\tau) = \frac{1}{2(N-2)\tau^2} \sum_{i=1}^{N-2} (X_{i+2} - 2X_{i+1} + X_i)^2 \tag{11}$$

N is the sampling sequence of gas concentration in time domain, τ is the sampling period, X_i is the sampling value. We measured the CO gas samples of 60 ppmv and calculated the Allan deviation of the acquisition data; the results are shown in Figure 6.

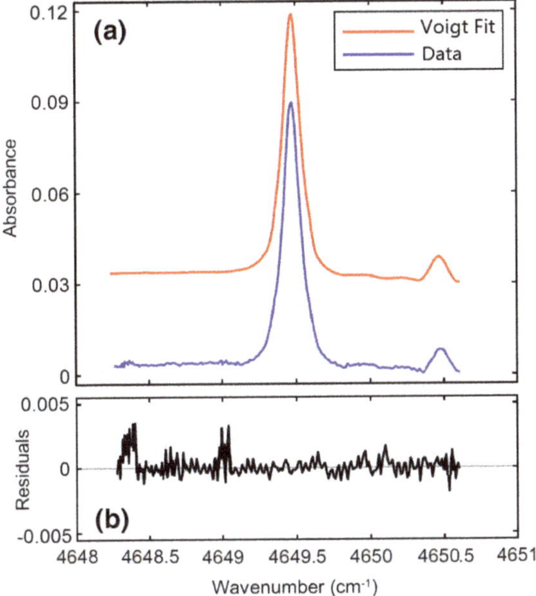

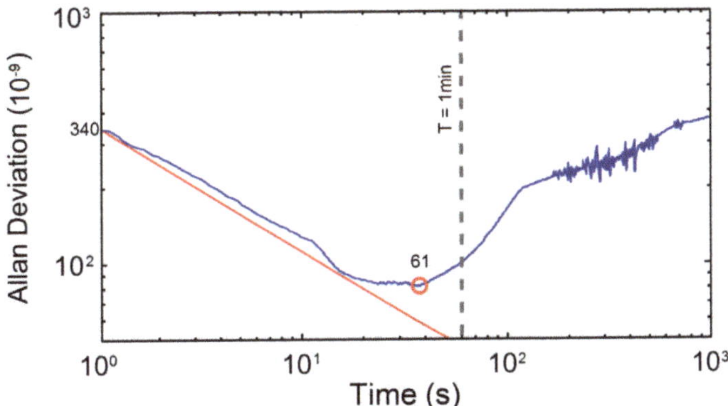

Figure 5. (**a**) Measured absorption spectrum (blue line) and Voigt theoretical spectrum (red line) via the LOP-DAST method. (**b**) Residual curve calculated by aforementioned two spectrums.

Figure 6. Allan deviation of the acquisition data of 60 ppmv CO.

It can be seen from the figure that the initial value of Allan deviation was about 340 ppbv, and the minimum value reached 61 ppbv with integration time of 40 s. This result is also similar to the value obtained from previous residual analyses.

5. Discussion and Conclusions

In this paper, a high-performance trace CO sensor using mid-infrared QCL combined with LOP-DAST is introduced. The working principle of LOP-DAST and calculating method of the absorption cross section are described, respectively. The stability of the sensor reached 1.7×10^{-2} by testing standard gas sample for a long period, and the MDL achieved approximate 100 ppbv by comparing the residual difference method and the Allan deviation method. Besides, the sensor is capable of detecting a variety of trace gases by changing the luminous source with different excitation wavelengths.

Comparing to state-of-the-art trace gas sensors using mid-infrared luminous source, simplified optical structure is utilized to achieve miniaturization and practicality, and maintain high measurement detection performance. Nevertheless, the shortcoming of the bulky package of the optical core devices and the auxiliary modules is obvious, such as bulky electrical power supply and Peltier temperature controller for QCL and HgCdTe detector, and multi-reflection cell with the size similar to a shoe box. Although the described sensor obtained ppbv levels of MDL, it was too large to be convenient for some special applications with compact volume requirement, such as gas detection in mobile and airborne platforms.

To meet the compact requirements of these specific applications, the challenge of fabrication technique on a chip level is of critical importance. Fortunately, the core components of the described mid-infrared gas sensor have been minimized, such as mid-infrared MEMS luminous source [22], micro-cavity absorption cell using silicon microring resonators [23,24], and on-chip HgCdTe photodiode detectors [25]. The advanced comments combined with integrated packaging technology to constitute new concept sensors on a chip level can potentially rival current infrared absorption spectrum sensors. Thus, micro mid-infrared gas sensors with ultra-compact size and high sensitivity appear to be very promising for future gas sensing.

Author Contributions: C.C. and Y.W. conceived the research and wrote the paper. Q.R., H.P., and P.W. performed the research.

Funding: This research was funded by the National Key R&D Program of China (2018YFC1503802, 2016YFC0303902), the National Natural Science Foundation of China (61871199), the Science and Technology Department of Jilin Province of China (20180201022GX), and the Education Department of Jilin Province of China (JJKH20180154KJ).

Conflicts of Interest: The authors declare no conflict of interest.

References

1. Stewart, G.; Mencaglia, A.; Philp, W. Interferometric signals in fiber optic methane sensors with wavelength modulation of the DFB laser source. *J. Lightw. Technol.* **1998**, *16*, 43–53. [CrossRef]
2. Wang, W.; Lv, Y. The principal, preparation and application of quantum cascade laser. *Laser J.* **2018**, *39*, 7–11.
3. Kourosh, K. *Sensors: An Introductory Course*; Springer Publications: New York, NY, USA, 2013; p. 29.
4. Chen, J.; Lin, H. The development of CH_4 detection technique. *Mod. Instrum.* **2007**, *13*, 1–3.
5. Mu, Y.; Hu, T.; Gong, H.; Ni, R.; Li, S. A Trace C_2H_2 Sensor Based on an Absorption Spectrum Technique Using a Mid-Infrared Interband Cascade Laser. *Micromachines* **2018**, *9*, 530. [CrossRef] [PubMed]
6. Li, M.; Bai, F. Design of high sensitivity infrared methane detector based on TDLAS-WMS. *Laser J.* **2018**, *39*, 75–79.
7. Mcmanus, J.B.; Shorter, J.H.; Nelson, D.D. Pulsed quantum cascade laser instrument with compact design for rapid, high sensitivity measurements of trace gases in air. *Appl. Phys. B* **2008**, *92*, 387–392. [CrossRef]
8. Gorrotxategi-Carbajo, P.; Fasci, E.; Ventrillard, I.; Carras, M.; Maisons, G.; Romanini, D. Optical-feedback cavity-enhanced absorption spectroscopy with a quantum-cascade laser yields the lowest formaldehyde detection limit. *Appl. Phys. B* **2013**, *110*, 309–314. [CrossRef]
9. Tang, Y.; Liu, W.; Kan, R.; Zhang, Y.; Liu, J.; Xu, Z.; Shu, X.; Zhang, S.; He, Y.; Geng, H.; et al. Spectroscopy processing for the NO measurement based on the room-temperature pulsed quantum cascade laser. *Acta Phys. Sin.* **2010**, *59*, 2364–2369.
10. Li, G.; Dong, M.; Song, N.; Song, F.; Zheng, C.; Wang, Y. Carbon monoxide gas detection system based on mid-infrared spectral absorption technique. *Spectrosc. Spectr. Anal.* **2014**, *34*, 2839–2844.
11. Wen, Z.; Wang, L.; Chen, G. development and application of quantum cascade laser based gas sensing system. *Spectrosc. Spectr. Anal.* **2010**, *30*, 2043–2048.
12. Li, Q.; Wang, Q.; Shang, T. Laser propulsion principle and development. *Laser Infrared* **2001**, *31*, 73–75.
13. Zhang, L.; Cui, X. The design of carbon monoxide detector based on tunable diode lasers absorption spectroscope. *Laser J.* **2014**, *35*, 54–56.

14. Sun, Y.; Liu, W.; Wang, S.; Huang, S.; Yu, X. Research on the method of interference correction for nondispersive infrared multi-component gas analysis. *Spectrosc. Spectr. Anal.* **2011**, *31*, 2719–2724.

15. Hao, N.; Zhou, B.; Chen, L. Measurement of nitrous acid and retrieval of aerosol parameters with differential optical absorption spectroscopy. *Acta Phys. Sin.* **2006**, *55*, 1529–1533.

16. Xu, J.; Xie, P.; Si, F.; Li, A.; Liu, W. Determination of tropospheric NO_2 by airborne multi axis differential optical absorption spectroscopy. *Acta Phys. Sin.* **2012**, *61*, 024204.

17. Chen, K.; Mei, M. Detection of gas concentrations based on wireless sensor and laser technology. *Laser J.* **2014**, *35*, 50–54.

18. Qiao, Y.; Tao, J.; Chen, C.-H.; Qiu, J.; Tian, Y.; Hong, X.; Wu, J. A miniature on-chip methane sensor based on an ultra-low loss waveguide and a micro-ring resonator filter. *Micromachines* **2017**, *8*, 160. [CrossRef]

19. Yin, Z.; Wu, C.; Gong, W.; Gong, Z.; Wang, Y. Voigt profile function and its maximum. *Acta Phys. Sin.* **2013**, *62*, 123301.

20. Werle, P.; Popov, P. Application of antimonide lasers for gas sensing in the 3–4-μm range. *Appl. Opt.* **1999**, *38*, 1494–1501. [CrossRef] [PubMed]

21. Ye, W.; Zheng, C.; Wang, Y. Stability measurement and temperature compensation of mid-infrared methane detection device. *Acta Opt. Sin.* **2014**, *34*, 0323003.

22. Lochbaum, A.; Fedoryshyn, Y.; Dorodnyy, A.; Koch, U.; Hafner, C.; Leuthold, J. On-chip narrowband thermal emitter for mid-IR optical gas sensing. *ACS Photonics* **2017**, *4*, 1371–1380. [CrossRef]

23. Khan, H.; Zavabeti, A.; Wang, Y.; Harrison, C.J.; Carey, B.J.; Mohiuddin, M.; Chrimes, A.F.; De Castro, I.A.; Zhang, B.Y.; Sabri, Y.M.; et al. Quasi physisorptive two dimensional tungsten oxide nanosheets with extraordinary sensitivity and selectivity to NO_2. *Nanoscale* **2017**, *9*, 19162–19175. [CrossRef] [PubMed]

24. Nitkowski, A.; Chen, L.; Lipson, M. Cavity-enhanced on-chip absorption spectroscopy using microring resonators. *Opt. Exp.* **2008**, *16*, 11930–11936. [CrossRef]

25. Sun, X.; Abshire, J.B.; Beck, J.D.; Mitra, P.; Reiff, K.; Yang, G. HgCdTe avalanche photodiode detectors for airborne and spaceborne lidar at infrared wavelengths. *Opt. Exp.* **2017**, *25*, 16589–16602. [CrossRef] [PubMed]

 *micromachines*

Article

A Trace C$_2$H$_2$ Sensor Based on an Absorption Spectrum Technique Using a Mid-Infrared Interband Cascade Laser

Ye Mu *, Tianli Hu, He Gong, Ruiwen Ni and Shijun Li

College of Information Technology, Jilin Agricultural University, Changchun 130018, China;
hutianli@jlau.edu.cn (T.H.); gonghe@jlau.edu.cn (H.G.); niruiwenjlau@163.com (R.N.); lishijun@jlau.edu.cn (S.L.)
* Correspondence: muye@jlau.edu.cn; Tel.: +86-130-8682-4573

Received: 12 August 2018; Accepted: 15 October 2018; Published: 19 October 2018

Abstract: In this study, tunable diode laser absorption spectroscopy (TDLAS) combined with wavelength modulation spectroscopy (WMS) was used to develop a trace C$_2$H$_2$ sensor based on the principle of gas absorption spectroscopy. The core of this sensor is an interband cascade laser that releases wavelength locks to the best absorption line of C$_2$H$_2$ at 3305 cm^{-1} (3026 nm) using a driving current and a working temperature control. As the detected result was influenced by $1/f$ noise caused by the laser or external environmental factors, the TDLAS-WMS technology was used to suppress the $1/f$ noise effectively, to obtain a better minimum detection limit (MDL) performance. The experimental results using C$_2$H$_2$ gas with five different concentrations show a good linear relationship between the peak value of the second harmonic signal and the gas concentration, with a linearity of 0.9987 and detection accuracy of 0.4%. In total, 1 ppmv of C$_2$H$_2$ gas sample was used for a 2 h observation experiment. The data show that the MDL is low as 1 ppbv at an integration time of 63 s. In addition, the sensor can be realized by changing the wavelength of the laser to detect a variety of gases, which shows the flexibility and practicability of the proposed sensor.

Keywords: trace C$_2$H$_2$ detection; mid-infrared spectrum; interband cascade laser; tunable semiconductor laser absorption spectroscopy; wavelength modulation technology; minimum detection limit

---

1. Introduction

Acetylene (C$_2$H$_2$) is one of the most important industrial gases used in industrial production, and it easily decomposes, burns, and explodes. Compared with other inflammable and explosive gases, C$_2$H$_2$ has a lower explosion limit. In recent years, there have been many reports of C$_2$H$_2$ explosion, which has brought great loss to people's safety and social production. Therefore, developing a sensor to monitor C$_2$H$_2$ with high accuracy and sensitivity in real time is highly important.

The absorption intensity of gas molecules in the mid-infrared band is nearly three orders of magnitude stronger than that in the visible or near-infrared spectrum band [1]. Under the same measurement conditions, the signal intensity obtained via gas concentration detection in the mid-infrared band is several times higher than that obtained in the visible and near infrared bands, which behaves with better detection accuracy and minimum detection limit (MDL) in the ppbv level [2–7]. Therefore, high sensitivity detection of essential C$_2$H$_2$ gas in chemical production using the spectrum absorption lines in mid-infrared band is an effective detection method [8,9].

Interband cascade lasers (ICLs), combined with tunable diode laser absorption spectroscopy (TDLAS)-wavelength modulation spectroscopy (WMS), are used to detect trace C$_2$H$_2$. The emitting light of ICL with a center wavelength of 3026 nm is tuned by the driving current and working temperature, which sweep the best absorption lines of C$_2$H$_2$. According to the Beer–Lambert law, the C$_2$H$_2$ concentration is deduced by measuring the attenuation of laser intensity. The proposed

C_2H_2 sensor's detection accuracy is 0.4%, the MDL is as low as 1 ppbv, and the stability is better than 1.776×10^{-2}.

2. Detection Principle of C_2H_2 Using Absorption Spectroscopy

2.1. Selection of C_2H_2 Absorption Line

The absorption spectrum refers to the fraction of incident radiation that is absorbed by the material over a range of frequencies. The absorption spectrum is primarily determined by the molecular composition of the material [10,11]. Except for diatomic and inert gases without polar symmetrical structure, each material has its own characteristic absorption spectrum. Thus, this characteristic can be used to identify gas molecules [12,13]. According to the high-resolution transmission molecular absorption database (HITRAN) database [14], the mid-infrared band absorption spectrum of C_2H_2 was searched to determine the absorption capacity of C_2H_2 at different wavelengths. As shown in Figure 1, C_2H_2 has a significant absorption spectrum line in the range of 3290–3320 cm^{-1} in comparison to H_2O, which may be present in significant amounts in the gas mixture.

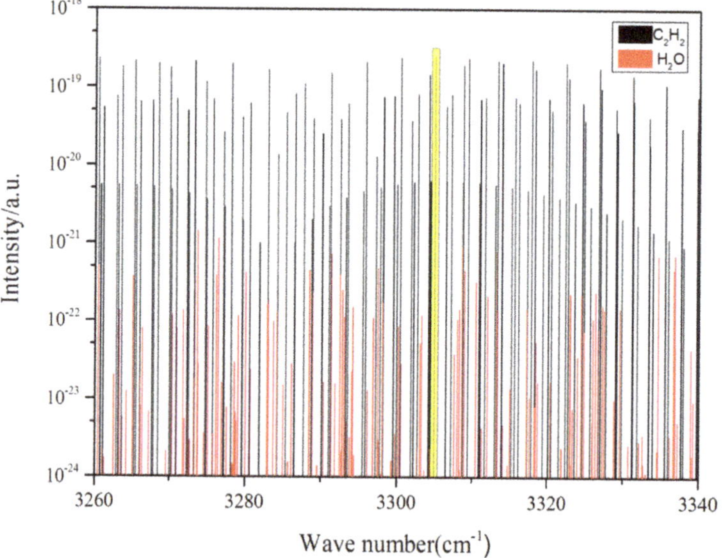

Figure 1. Absorption lines of C_2H_2 and H_2O in the range of 3290–3320 cm^{-1}.

As shown in Figure 1, the x coordinate is the wave number (reciprocal centimeters, cm^{-1}), the y coordinate is the absorption intensity (a.u.), the black lines represent the C_2H_2 gas absorption lines, and blue lines are H_2O absorption lines. To improve the accuracy of the measured results, a C_2H_2 absorption line centered at 3305 cm^{-1} (3026 nm) with an absorption magnitude of 10^{-19} was selected as the optimum C_2H_2 target line. All of the H_2O absorption lines (the two closest lines are located at 3303 cm^{-1} and 3308 cm^{-1}) under an absolute humidity of 2% did not interfere with the selected C_2H_2 line at 3305 cm^{-1}, since they were $\sim$2 to 3 cm^{-1} away. With a higher relative humidity, the dryers could be used to lower the H_2O concentration and thereby reduce the absolute humidity to an acceptable level, e.g., below 2%. In that case, the sensor could operate normally.

2.2. Derivation of TDLAS-WMS

TDLAS is based on the principle of the Beer–Lambert law, which states that absorbance is proportional to the concentrations of the attenuating species in the material sample [15,16]. It can be expressed as follows:

$$I_1 = I_0 e^{-\alpha(v)PCL} \tag{1}$$

where I_0 is the emitting light intensity of the laser, I_1 is the light intensity after passing the measured gas, L is the effective length of absorption optical path, P is the pressure in the cell, C is the gas concentration, and $\alpha(v)$ is the molecular absorption coefficient. Then, $\alpha(v)$ can be expressed as follows:

$$\alpha(v) = T(t) \times g(v - v') \times N \tag{2}$$

where $T(t)$ is the absorption intensity of gas at time point t, $g(v - v')$ is a linear function of the measured gas, v' is the initial frequency of energy level transition of the gas molecule, and N is the number of molecules per volume. To improve the MDL performance of the C_2H_2 sensor, TDLAS-WMS was adopted to eliminate the $1/f$ noise caused by ICL or external environmental disturbances [17]. The time dependent wavelength of the ICL can be described as [17–19]:

$$v_1(t) = v_0(t) + A\cos(\omega t) \tag{3}$$

where $v_0(t)$ is the central frequency of emitting light, which is determined by the low-frequency component of driving signal, and A and ω are the amplitude and frequency of the high-frequency component of the driving signal, respectively. By substituting Formula (3) into Formula (1) and expanding it in the form of cosine Fourier series:

$$v_1(t) = v_0(t) + A\cos(\omega t) \tag{4}$$

where A_n is the amplitude of each harmonic component and can be expressed as follows [20]:

$$A_n(v_0) = \frac{I_0 \times 2^{1-n} \times C \times L}{n!} \times A^n \times \left. \frac{d^n \alpha}{dv^n} \right|_{v = v_0} \tag{5}$$

According to Formula (5), the amplitude of the first harmonic component is:

$$A_1(v_0) = I_0 L A \frac{d\alpha}{dv} |_{v=v_0} \tag{6}$$

The amplitude of the second harmonic component is:

$$A_2(v_0) = \frac{I_0 C L}{4} A^2 \left. \frac{d^2\alpha}{dv^2} \right|_{v = v_0} \tag{7}$$

Based on the above formulas, the amplitudes of the odd harmonic components at the center frequency were 0, and the even harmonic components at the center frequency reached maximum values, which were positively proportional to the gas concentration. As the order increased, the amplitude decreased gradually. In summary, TDLAS-WMS is the optimum choice to analyze the measured gas concentration, which can effectively reduce the $1/f$ noise, increase the signal-to-noise ratio, and improve the MDL performance of the sensor [17].

3. System Configuration

The ICL laser produced by Nanoplus Co., Gerbrunn, Germany, was used as the light source. Its output wavelength is in the range of 3023 nm to 3027 nm, and the central wavelength is 3025 nm.

The embedded thermoelectric cooler Peltier was combined with negative temperature feedback control to guarantee the stability of working temperature of ICL during operation. In terms of the multi-reflection gas cell, the physical length was 40 cm with a volume of 500 mL. The laser was reflected 52 times in the cell, and the effective optical length was increased by up to 20 m. The photodetector is a mid-infrared photoelectric detector: VL5T0 produced by Thorlabs. The detector has good linearity in the spectrum range of 2.7 μm to 4.5 μm, and the response time is less than 120 ns. In addition, the Signal Recovery 7280 lock-in amplifier (LIA) was used to demodulate the second harmonic signal. The trace C_2H_2 detector was mainly divided into two modules: electrical and optical. The overall schematic diagram is shown in Figure 2.

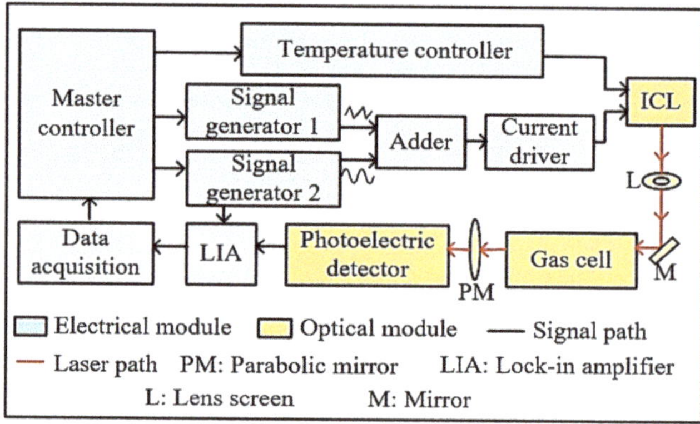

Figure 2. Schematic diagram of the C_2H_2 detector.

The master controller controls signal generator 1, which generates a sinusoidal wave signal with 5 kHz frequency and 0.026 V amplitude, and signal generator 2, which generates a triangular wave signal with 0.5 Hz frequency and 0.2 V amplitude. A high-frequency sinusoidal signal is transmitted to the phase lock-in amplifier as a reference signal. Besides, it is added with a low-frequency triangular wave signal to drive the ICL laser. The emitting light of the ICL with a center wavelength of 3026 nm is tuned by the driving current and the working temperature, which converges through the aperture (L) and goes into the gas cell reflected by the lens (M). Through the measured C_2H_2 gas, the output beam is transformed into an electric signal by a photoelectric detector, and then transmitted to the phase locked amplifier. Signal generator 2 provides the phase-locked amplifier with a synchronous signal to ensure phase synchronization. At the output terminal of the phase-locked amplifier, the second harmonic signal can be obtained. Finally, the data acquisition unit processes the measured gas concentration.

4. Experiment

4.1. Response

To observe the working performance of the trace C_2H_2 gas detector, five C_2H_2 gases with different standard concentrations (20, 40, 60, 80, and 100 ppbv) were prepared using a dynamic gas dilution equipment. The prepared C_2H_2 gases were pumped into the gas cell in sequence at 5 min intervals, and the corresponding peak voltages of the second harmonic signal were obtained and denoted as $max(2f)$.

As shown in Figure 3, the x coordinate is the measured time, and the y coordinate is the peak of the second harmonic signal. By analyzing the absorption of the emitting light power from the C_2H_2 gas, the peak of the second harmonic signal was linearly decreased by the C_2H_2 gas concentration. As a result, this peak was used to represent the C_2H_2 gas concentration.

Nevertheless, due to fluctuations of ICL output power, the value of $max(2f)$ changed slowly over a long observation time (longer than 1 h), and the measured results of different concentrations showed the same growth trend. Seen from Formulas (6) and (7), the laser-induced intensity I_0 as a critical factor of system long-term drift is contained in both of them. The ratio of the second harmonic component to first harmonic component, named as the $2f/1f$-WMS technique, can be utilized to reduce the impact caused by the ICL output power fluctuations [21].

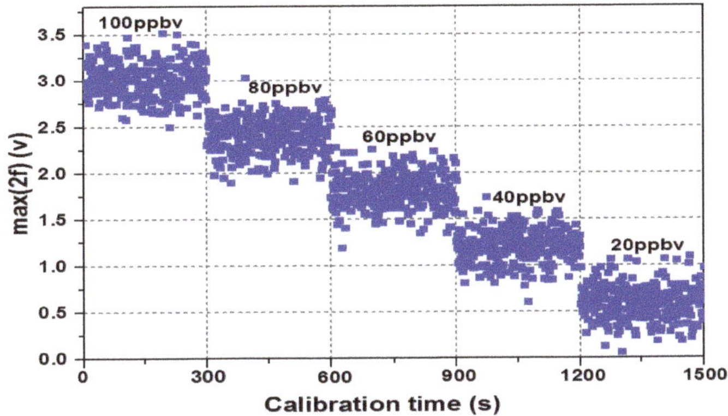

Figure 3. *Max(2f)* at different concentrations.

4.2. Precision

Detection precision is a critical parameter for evaluating the sensor performance. The relationship curves of *max(2f)* between standard concentration and the measured concentration are shown in Figure 4.

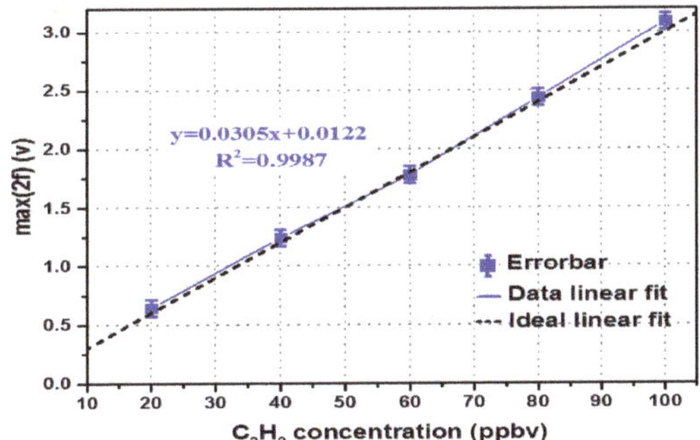

Figure 4. Relationship curves of *max(2f)* between the standard concentration and the measured concentration.

As shown in Figure 4, the x coordinate is the standard concentration of C_2H_2, and the y coordinate is the measured peak value of the second harmonic signal. The mean value of each data is expressed in the form of error bars, and the solid blue line is the relationship between the average voltage value of the measured data and the concentration of C_2H_2 gas. The black dotted line shows the relationship between the C_2H_2 gas concentration and the peak value of the second harmonic signal under theoretical conditions. The results showed that the maximum deviation of the measured data is 0.0412 V, and that the accuracy is 0.4%.

Formula (8) can be obtained by linearly fitting the measured data:

$$max(2f) = 0.0305C + 0.0122 \tag{8}$$

where C (parts-per-billion volume, ppbv) is the concentration of C_2H_2, and then:

$$C = 32.7869 \times max(2f) - 0.4 \qquad (9)$$

Formula (9) can be used to convert the measured peak value of the second harmonic signal into the corresponding C_2H_2 gas concentration.

4.3. Stability

The stability deals with the degree to which sensor characteristics remain constant over time, which is determined by computing the ratio of the maximum deviation and the mean value for a long time of observation at a specific concentration of C_2H_2 gas [22]. At room temperature, the C_2H_2 gas with a concentration of 1 ppmv was observed for 2 h, the relationship between the gas concentrations was detected using the proposed sensor, and the detection time was recorded. The results are shown in Figure 5.

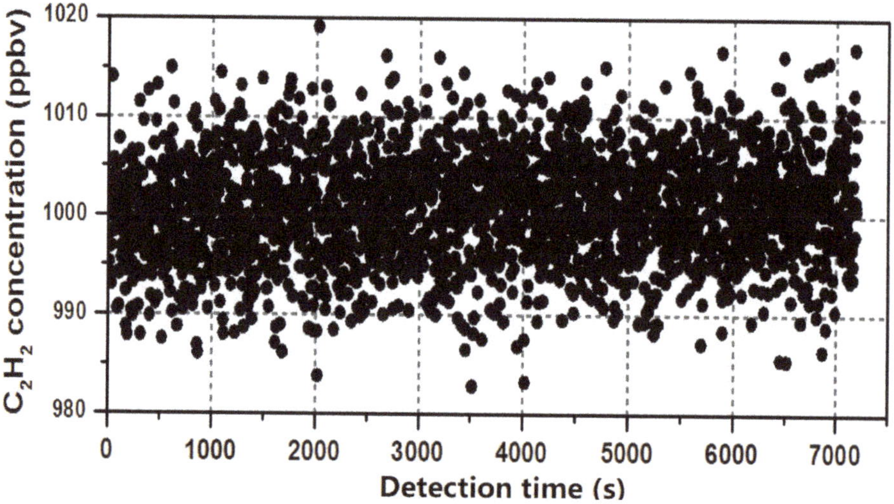

Figure 5. Detection results of 1 ppmv C_2H_2 gas concentration in 2 h.

The x coordinate is the detection time, and the y coordinate is the measured concentration. During the 2 h experimental observation, the peak value of the second harmonic signal ranged from 980 ppbv to 1020 ppbv, and more than 90% of the results were in the range of 990 ppbv to 1010 ppbv with $\pm$ 10 ppbv fluctuation. The mean of the measured results was 1000.52 ppbv, and the maximum deviation of the actual data was 17.76 ppbv. Thus, the stability was better than 1.776×10^{-2}.

In addition, the experimental results changed slowly with the ppbv level during the 2 h observation because of the drift noise of the proposed sensor, indicating that the measurement precision was a critical factor in long-term observation. A reference cell fully filled with pure nitrogen could be utilized to suppress the long-term common noise in future work.

4.4. MDL

As the measured output data drift with time when detecting gas concentration, Allan variance [23] was used to evaluate the experimental data of 1 ppmv C_2H_2, as shown in Figure 6.

The results were carried out under laboratory conditions, and the system sampling rate was 10 Hz. As illustrated in Figure 6, the obtained MDL of the proposed sensor was 30 ppbv, with an integration time of 0.1 s. The results of Allan variance analysis showed an appropriate integration time of 63 s, corresponding to an MDL of ~0.958 ppbv. In addition, white noise, one of the main sensor noises, is a random signal having equal intensity at different frequencies. The decreasing red solid line, which is about −1/2, indicates that the theoretical expected behavior of a system is dominated by white noise (before 63 s) [21]. The MDL began to increase after an integration time of 63 s, because the system drift noise dominated in this area.

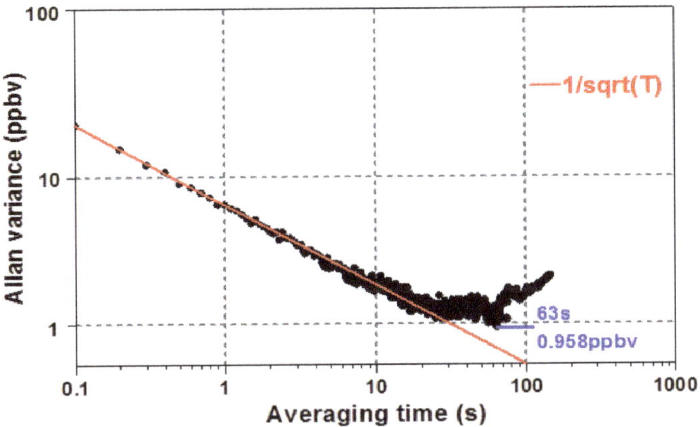

Figure 6. Allan variance of 1 ppmv C_2H_2.

4.5. Recovery Time and Reproducibility

Two C_2H_2 samples with different concentration levels of 0 ppmv (Pure Nitrogen 99.999%) and 1 ppmv, generated by dynamic gas dilution equipment, were measured to test the performance of the recovery time and reproducibility of the proposed sensor, and the dynamic measured results are shown in Figure 7.

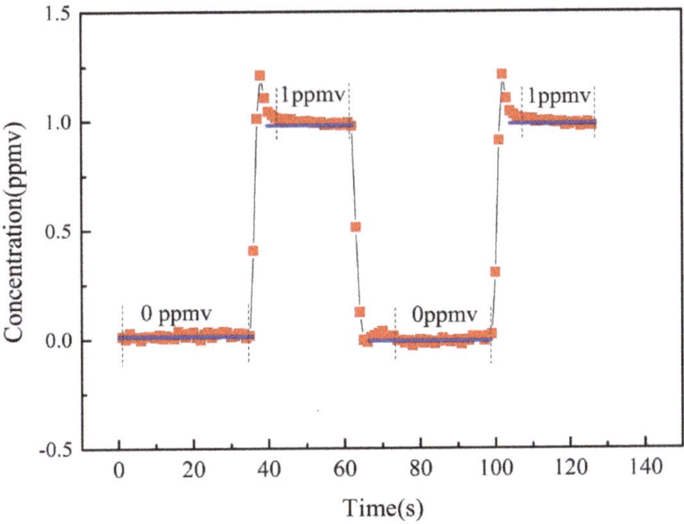

Figure 7. Dynamic measured results of C_2H_2 sensor.

The total measurement time is 130 s under a pressure condition of 760 torr. To test the reproducibility performance, the C_2H_2 concentration was initially changed from 0 ppmv to 1 ppmv, then it decreased to 0 ppmv, and finally increases to 1 ppmv. The recovery time includes the gas distribution time, and the process of gas preparation is related to PID control algorithm utilized by dynamic gas dilution equipment. The recovery time for gas sample preparation from a low concentration to high concentration is longer than from high concentration to low concentration, 10 s (0 to 1 ppmv), 15 s (1 to 0 ppmv).

4.6. Performance Comparison

In recent years, many researchers have conducted in-depth studies on the detection of C_2H_2 gas. The C_2H_2 sensor developed in this study is compared with the reported C_2H_2 sensors, as shown in Table 1.

Table 1. Performance comparison of the proposed C_2H_2 sensor and the reported C_2H_2 sensors.

Ref/Type	Wavelength/Maximum Intensity	Technique	MDL (ppmv)	Error (%)
[24]	1.533 μm/1.211×10^{-20}	DAS	1.8	4
[25]	1.534 nm/8.572×10^{-21}	TDLAS-WMS	2	1
LGA-4500	1.533 μm/1.211×10^{-20}	TDLAS-WMS	0.1	1
[26]	1.523 μm/3.145×10^{-20}	CRDS	0.00034	/
[27]	Broad mid-infrared range	CEAS	0.5	/
This study	1.533 μm/1.211×10^{-20}	TDLAS-WMS	0.001	0.4

Both sensors in [24,25] and the LGA-4500 C_2H_2 sensor can detect the gas concentration using the near-infrared band, where the absorption intensity of C_2H_2 ranges from 10^{-21} to 10^{-20}. Because of the absorption of C_2H_2 in the near-infrared band, which is three orders of magnitude weaker than that in the mid-infrared band, the MDL of C_2H_2 sensor using absorption line in the near-infrared band remains at the order of ppmv. In addition, because of the use of traditional direct absorption spectroscopy (DAS) in [24], the MDL is still far behind that of the LGA-4500 sensor, although they both use the same absorption band. Therefore, TDLAS-WMS has better MDL performance compared with traditional DAS. In this study, the proposed C_2H_2 sensor uses strong absorption line in the mid-infrared band and high-sensitivity TDLAS to obtain superior MDL.

To improve the poor MDL caused by weak absorption intensity in the near-infrared band, one of the most common cavity-enhanced absorption techniques, cavity ring-down spectroscopy (CRDS), is utilized [26]. The MDL of 340 pptv is achieved by employing an external optical cavity with high-reflectivity mirrors as a sample cell. Compared with TDLAS-WMS, CRDS has some disadvantages. First, spectra data cannot be acquired rapidly due to the monochromatic laser source, and the response time is usually at the minute level. Second, analysis is limited by the availability of the tunable laser light at the appropriate wavelength and also at the availability of high-reflectivity mirrors at these wavelengths. Finally, the requirement for laser systems and high-reflectivity mirrors often makes CRDS more expensive than TDLAS-WMS.

Another different spectroscopic method has been developed to obtain gas concentrations with high sensitivity: cavity-enhanced absorption spectroscopy (CEAS). Taking advantage of the high spatial coherence and high brightness of the broadband supercontinuum source, methane and C_2H_2 are detected using a mid-infrared spectrum over a bandwidth as large as 450 nm [27]. A MDL of 0.5 ppm for C_2H_2 and 0.25 ppm for methane is measured simultaneously, according to the linear response function. Although this gas sensor prototype can retrieve gas concentrations with sub-ppm levels, the MDL and measurement speed of the CEAS technique should be improved in further studies. First, the power spectral density of the mid-infrared source coupled into the chamber can be increased by reducing the connection losses between the light source and the fiber. Second, the measurement speed is currently limited to 30 nm/min, due to monochromator scanning and the long integration time that is needed to improve the signal-to-noise ratio. Although this could be significantly reduced by using a spectrometer with a detector array, a detector with high sensitivity is generally required, which makes this technique more expensive and impractical.

5. Conclusions

A trace C_2H_2 gas sensor was developed using TDLAS-WMS with an absorption spectrum line at $3305\ cm^{-1}$ (3026 nm). The sensor included an ICL laser, a gas chamber with a 20 m-long optical path, a photodetector, and a phase-locked amplifier. The detection results of the C_2H_2 gas with five different concentrations showed a good linear relationship between the peak value of the second harmonic signal and the gas concentration, with a linearity of 0.9987 and a detection accuracy of 0.4%. In total, 1 ppmv of C_2H_2 gas sample was used for a 2 h observation, and the measured data show the MDL is as low as 1 ppbv at an integration time of 63 s. In addition, the sensor can be realized by changing the wavelength of the laser to detect a variety of gases, demonstrating flexibility and practicability.

Although the proposed sensor achieved ppbv scales of MDL, it was too large to be suitable and convenient for some applications, such as field measurements (e.g., mobile and airborne). Our new motivation is to develop a gas sensor that is compact and rugged, because mechanical fiber coupling of the diode lasers did not need to be adjusted over several months. Fiber delivery and a fiber beam-coupler will be utilized in our future design, which can reduce the size and thus the ease of operation.

Author Contributions: Y.M., T.H., and C.C. conceived and designed the research. Y.M., T.H., C.C., H.G., R.N., and S.L. performed the research. Y.M., T.H., and C.C. wrote the paper.

Acknowledgments: This research was supported by Science and Technology Department of Jilin Province, China (20160623016TC, 20170204017NY, 20170204038NY, 20180201022GX), College Students Innovation and Entrepreneurship Training Program of Jilin Province, China (2017490, 2017493).

Conflicts of Interest: The authors declare no conflict of interest.

References

1. Chong, X.; Li, E.; Squire, K.; Wang, A.X. On-chip near-infrared spectroscopy of CO_2 using high resolution plasmonic filter array. *Appl. Phys. Lett.* **2016**, *108*, 221106. [CrossRef]
2. Chen, B.; Han, C.; Liu, G. Detection on Particulate Pollutantin Transformer Oil Based on the Mid-Infrared Spectrum. *Acta Photonica Sin.* **2016**, *45*, 530002. [CrossRef]
3. Zhang, L.; Cui, X. The Design of Carbon Monoxide Detector Based on Tunable Diode Lasers Absorption Spectroscope. *Laser J.* **2014**, *35*, 54–56.
4. Xie, Y.; Li, F.; Fan, X.; Hu, S.; Xiao, X.; Wang, J. Components Analysis of Biochar Based on Near Infrared Spectroscopy Technology. *Chin. J. Anal. Chem.* **2018**, *46*, 609–615. [CrossRef]
5. Li, M.; Bai, F. Design of High Sensitivity Infrared Methane Detector Based on TDLAS-WMS. *Laser J.* **2018**, *39*, 75–79.
6. Northern, H.; O'Hagan, S.; Hamilton, M.L.; Ewart, P. Mid-infrared multi-mode absorption spectroscopy, MUMAS, using difference frequency generation. *Appl. Phys. B Lasers Opt.* **2015**, *118*, 343–351. [CrossRef]
7. Chen, C.; Wang, B.; Li, C.; Li, J.; Wang, Y. A Trace Gas Sensor Using Mid-infrared Quantum Cascaded Laser at 4.8 μm to Detect Carbon Monoxide. *Spectrosc. Spectr. Anal.* **2014**, *34*, 838–842.
8. Yan, M.; Luo, P.; Iwakuni, K.; Millot, G.; Hänsch, T.; Picqué, N. Mid-infrared dual-comb spectroscopy with electro-optic modulators. *Light Sci. Appl.* **2017**, *6*, e17076. [CrossRef] [PubMed]
9. Sun, Z.; Li, Z.; Li, B.; Alwahabi, Z.; Aldén, M. Quantitative C_2H_2 Measurements in Sooty Flames Using Mid-Infrared Polarization Spectroscopy. *Appl. Phys. B Lasers Opt.* **2010**, *101*, 423–432. [CrossRef]
10. Chen, K.; Mei, M. Detection of Gas Concentrations Based on Wireless Sensor and Laser Technology. *Laser J.* **2014**, *35*, 50–54.
11. KC, U.; Nasir, E.; Farooq, A. A mid-infrared absorption diagnostic for acetylene detection. *Appl. Phys. B Lasers Opt.* **2015**, *120*, 223–232. [CrossRef]
12. Van Helden, J.H.; Lang, N.; Macherius, U.; Zimmermann, H.; Ropcke, J. Sensitive trace gas detection with cavity enhanced absorption spectroscopy using a continuous wave external-cavity quantum cascade laser. *Appl. Phys. Lett.* **2013**, *103*, 553. [CrossRef]

13. Jin, M.; Lu, F.; Belkin, M. High-sensitivity infrared vibrational nanospectroscopy in water. *Light Sci. Appl.* **2017**, *6*, e17096. [CrossRef] [PubMed]

14. Rothman, L.S.; Gamache, R.R.; Goldman, A.; Brown, L.R.; Toth, R.A.; Pickett, H.M.; Poynter, R.; Flaud, J.M.; Camy-Peyret, C.; Barbe, A.; et al. The HITRAN database: 1986 edition. *Appl. Opt.* **1987**, *26*, 4058–4097. [CrossRef] [PubMed]

15. Li, J.; Yu, B.; Zhao, W.; Chen, W. A Review of Signal Enhancement and Noise Reduction Techniques for Tunable Diode Laser Absorption Spectroscopy. *Appl. Spectrosc. Rev.* **2014**, *49*, 666–691. [CrossRef]

16. Zhu, X.; Lu, W.; Rao, Y.; Li, Y.; Lu, Z.; Yao, S. Selection of baseline method in TDLAS direct absorption CO_2 measurement. *Chin. Opt.* **2017**, *10*, 455–461.

17. Klein, A.; Witzel, O.; Ebert, V. Rapid, Time-Division Multiplexed, Direct Absorption- and Wavelength Modulation-Spectroscopy. *Sensors* **2014**, *14*, 21497–21513. [CrossRef] [PubMed]

18. Gao, G.; Chen, B.; Hu, B. A system for gas sensing employing correlation spectroscopy and wavelength modulation techniques with a multimode diode laser. *Measurement* **2013**, *46*, 1657–1662. [CrossRef]

19. Qu, S.; Wang, M.; Li, N. Mid-Infrared Trace CH_4 Detector Based on TDLAS-WMS. *Spectrosc. Spectr. Anal.* **2016**, *36*, 3174–3178.

20. Wang, M.; Zhang, Y.; Liu, W.; Liu, J.; Wang, T.; Tu, X.; Gao, S.; Kan, R. Second-harmonic Detection Research with Tunable Diode Laser Absorption Spectroscope. *Opt. Tech.* **2005**, *31*, 279–285.

21. Li, C.; Dong, L.; Zheng, C.; Tittel, F.K. Compact TDLAS based optical sensor for ppb-level ethane detection by use of a 3.34 μm room-temperature CW interband cascade laser. *Sens. Actuators B Chem.* **2016**, *232*, 188–194. [CrossRef]

22. Jacob, F. *Handbook of Modern Sensors*, 4th ed.; Springer: New York, NY, USA, 2010; pp. 13–52. ISBN 978-1-4419-6465-6.

23. Allan, D.W.; Barnes, J.A. A modified "Allan Variance" with increased oscillator characterization ability. In Proceedings of the 35th Annual Symposium on Frequency Control, Philadelphia, PA, USA, 27–29 May 1981; pp. 470–475.

24. Deng, H. Research on Near Infrared Absorption Spectrum Theory and Detection Technology for Acetylene. Master's Thesis, Anhui University, Hefei, China, 2017.

25. He, Q.; Liu, H.; Li, B.; Pan, J.; Wang, L.; Zheng, C.; Wang, Y. Online Detection System on Acetylene with Tunable Diode Laser Absorption Spectroscopy Method. *Spectrosc. Spectr. Anal.* **2016**, *36*, 3501–3505.

26. Schmidt, F.M.; Vaittinen, O.; Metsälä, M.; Kraus, P.; Halonen, L. Direct detection of acetylene in air by continuous wave cavityring-down spectroscopy. *Appl. Phys. B Lasers Opt.* **2010**, *101*, 671–682. [CrossRef]

27. Amiot, C.; Aalto, A.; Ryczkowski, P.; Toivonen, J.; Genty, G. Cavity enhanced absorption spectroscopy in the mid-infrared using a supercontinuum source. *Appl. Phys. Lett.* **2017**, *111*, 061103. [CrossRef]

 *micromachines* 

Article

Midwave FTIR-Based Remote Surface Temperature Estimation Using a Deep Convolutional Neural Network in a Dynamic Weather Environment

Sungho Kim [1,*], **Jungho Kim** [2], **Jinyong Lee** [2] **and Junmo Ahn** [2]

[1] Department of Electronic Engineering, Yeungnam University, 280 Daehak-Ro, Gyeongsan, Gyeongbuk 38541, Korea
[2] Agency for Defense Development, 111 Sunam-dong, Daejeon 34186, Korea; jhkim@add.re.kr (J.K.); jinylee@add.re.kr (J.L.); ahnjm@add.re.kr (J.A.)
* Correspondence: sunghokim@ynu.ac.kr; Tel.: +82-53-810-3530

Received: 4 September 2018; Accepted: 26 September 2018; Published: 27 September 2018

Abstract: Remote measurements of thermal radiation are very important for analyzing the solar effect in various environments. This paper presents a novel real-time remote temperature estimation method by applying a deep learning-based regression method to midwave infrared hyperspectral images. A conventional remote temperature estimation using only one channel or multiple channels cannot provide a reliable temperature in dynamic weather environments because of the unknown atmospheric transmissivities. This paper solves the issue (real-time remote temperature measurement with high accuracy) with the proposed surface temperature-deep convolutional neural network (ST-DCNN) and a hyperspectral thermal camera (TELOPS HYPER-CAM MWE). The 27-layer ST-DCNN regressor can learn and predict the underlying temperatures from 75 spectral channels. Midwave infrared hyperspectral image data of a remote object were acquired three times a day (10:00, 13:00, 15:00) for 7 months to consider the dynamic weather variations. The experimental results validate the feasibility of the novel remote temperature estimation method in real-world dynamic environments. In addition, the thermal stealth properties of two types of paint were demonstrated by the proposed ST-DCNN as a real-world application.

Keywords: midwave infrared; thermal radiation; hyperspectral; remote surface temperature; weather variation; deep learning; regressor; thermal stealth

---

1. Introduction

The relationships between solar radiance and emitted thermal radiative energy are important for infrared stealth technology. Radiated solar energy (6000 K) heats an object, which then radiates thermal energy according to Planck's law [1]. One of the core technologies in infrared stealth research is to measure the surface temperature of a remote object.

The surface temperature of an object is independent of the wavelength, and can be estimated from even a single spectral band with a known atmospheric transmissivity and surface emissivity. In addition, the atmospheric conditions are crucial for surface temperature estimation [2]. The atmospheric weather conditions (e.g., temperature, humidity) change dynamically on earth, which leads to wide variations of spectral transmissivity. An incorrect atmospheric transmissivity hinders remote temperature estimation. Many studies have been conducted to measure remote temperature as correctly as possible.

Conventional approaches usually require an atmospheric correction to remotely estimate the temperature [3]. A previous study proposed a temperature estimation using single-channel infrared information and known atmospheric transmissivity [4]. A single band-based temperature emissivity separation (TES) was applied to estimate absolute land surface temperature from the MODerate resolution atmospheric TRANsmission (MODTRAN) -based atmospheric transmissivity information [2]. A single channel-based temperature estimation is impractical because of the inaccurate atmospheric profile information.

A two-channel-based method such as a split window algorithm can estimate the surface temperature as a linear function of two brightness temperatures [5]. Although the split-window method does not require information on the atmospheric transmissivity at the time of the temperature measurement, it requires accurate differential water vapor absorption in two adjacent thermal infrared channels [6].

A four channel-based surface temperature estimation method was proposed in Reference [7]. They reported an improved surface temperature estimation using three longwave thermal channels (8.7, 10.8, 12.0 μm) and one midwave thermal channel (3.9 μm) compared to the split window methods.

A multi-channel method was proposed using longwave hyperspectral thermal infrared for high-emissivity surfaces [8]. They used ten manually selected channels and applied a least square minimization to estimate the parameters. The multi-channel method [9] was improved by considering thirty-six channels and unknown emissivity in a linear system [8].

An accurate remote surface temperature estimation is difficult under dynamically varying weather conditions. The above-mentioned approaches (1-channel, 2-channel, 4-channel, and multi-channel) have their own advantages and disadvantages. These methods work if the specific conditions are satisfied, such as the known atmospheric transmissivity, known water vapor contents, and known surface emissivity. On the other hand, they cannot guarantee the temperature accuracy if the required atmospheric information is unavailable online or the weather conditions change abruptly.

In this paper, the problem of a remote surface temperature estimation in a dynamic weather environment is solved by focusing on the sensor, database (DB), and deep learning scheme. A new hyperspectral thermal infrared camera (HYPER-CAM MWE, TELOPS, Quebec, QC, Canada) was adopted to analyze both the solar radiance and thermal radiation in the 1.5–5.5 μm band. This hyperspectral camera can provide 374 spectral bands with a calibrated spectral radiance. A dynamic weather database was recorded three times a day (10:00, 13:00, and 15:00) for 7 months to cover a wide range of atmospheric variations in a coastal environment. The proposed surface temperature-deep convolutional neural network (ST-DCNN) can estimate the temperatures by learning the network on a huge spectral DB.

The remainder of this paper is organized as follows. Section 2 introduces the background of Fourier transform infrared (FTIR)-based brightness temperature measurement process. Section 3 explains the overall structure of the paper, including the hyperspectral database construction method and deep convolutional neural network-based temperature estimation with the ST-DCNN. Section 4 evaluates the remote surface temperature estimation performance of the proposed method by comparing it with the baseline methods. The paper is concluded in Section 5.

2. Background: Passive Open Path Fourier Transform Infrared (OP-FTIR)

The remote surface temperature estimation was based on the brightness temperature as shown in Figure 1. The research objective was to estimate the temperature of a surface heated by solar energy. The surface heated by direct sunlight was rotated 180° after 20 min to determine the thermal radiation. The thermal energy radiated by the rotated surface passes through the atmosphere, which alters the spectral radiation. The atmospheric transmittance changes according to the molecular contents, such as carbon dioxide and water vapor. Additional thermal energy coming from the atmosphere was added to the remote surface radiation. This thermal radiation was recorded in a Michelson interferometer and the spectra were obtained by applying the Fourier transform to the interferograms. The brightness temperatures were obtained through the spectral-radiometric calibration and inverse of Planck's law.

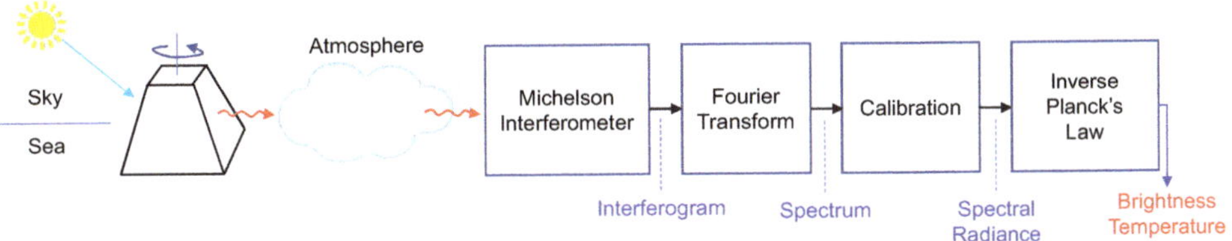

Figure 1. Overall process of the brightness temperature extraction using the passive open path Fourier transform infrared (FTIR) imaging system.

2.1. Passive Open Path Michelson Interferometer

An open path Michelson interferometer can be prepared using an active or passive approach for outdoor applications [10]. The former uses an active IR source that usually heats between 1000 and 1800 °C. The maximum range for an active source open path Fourier transform infrared (OP-FTIR) system is approximately 500 m. On the other hand, the latter has no sending unit. The infrared source is generally the sun (absorption) or a preheated object surface (emission). In the sun source, solar energy passes through gaseous material and the amount of absorbed energy can be measured. Figure 2 presents the basic diagram of a passive open path Michelson interferometer. Although the sensitivity of the passive OP-FTIR is generally less than that of the active method, the effective range is longer (i.e., up to several kilometers), which is suitable for remote sensing applications.

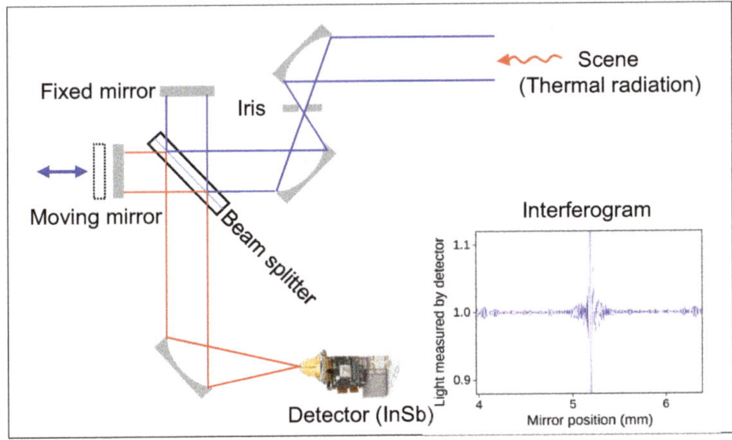

Figure 2. Basic diagram of a Michelson interferometer for passive open path applications.

A Michelson interferometer receives an input beam of radiation, divides the beam into two paths, and then recombines the two beams after a path difference [11]. A beam splitter divides the beam into two paths, and two mirrors, where one of the mirrors is movable, can make a path difference after the beams reflect off the mirrors. The recombined two beams at the beam splitter generate an interference pattern that is recorded in the detector (indium antimonide, InSb). This interference pattern is called an interferogram, which is the raw data from a passive OP-FTIR sensor. Figure 3a presents an interferogram image at zero path difference (ZPD), whose equivalent optical path difference (OPD) ID is 593. Figure 3b gives an example of the whole interferogram at pixel (190, 10). The following subsections outline the mathematical formulations of how to convert raw interferograms to a brightness temperature [12].

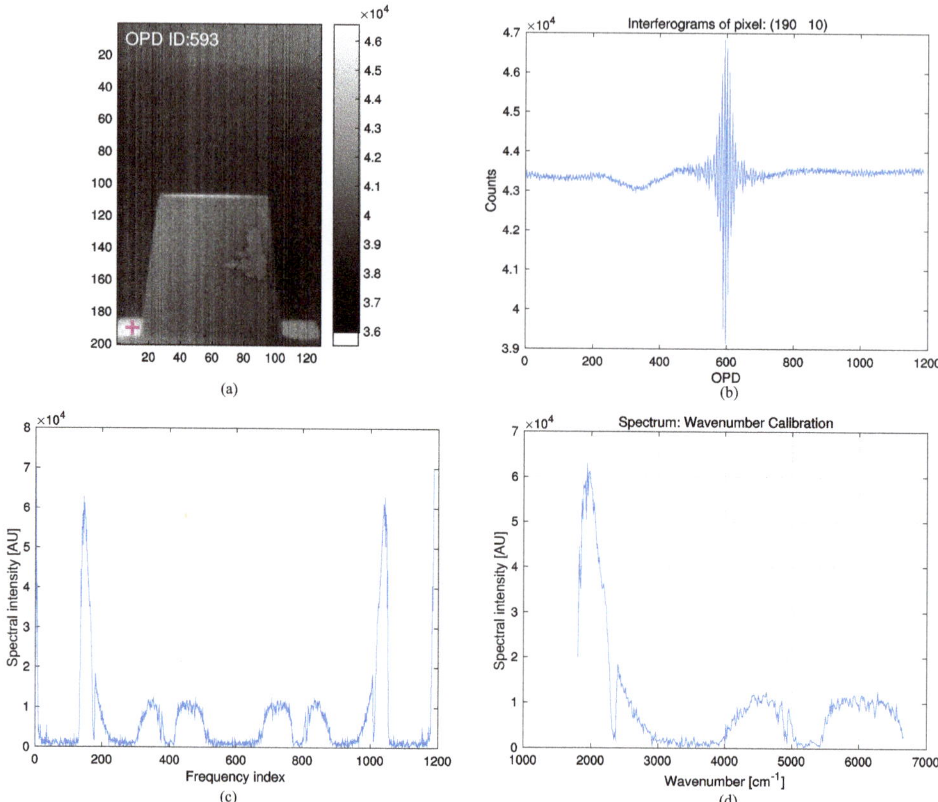

Figure 3. Spectral calibration process: (**a**) interferogram image at optical path difference (OPD) ID = 592 (ZPD); (**b**) interferogram at a pixel ((row, col) = (190, 10)); (**c**) fast Fourier transform (FFT) results; (**d**) wavenumber calibration results.

2.2. Fourier Transform

The intensity as a function of the path difference in the interferometer p and wavenumber $\widetilde{\nu} = 1/\lambda$ is expressed as Equation (1) [13]:

$$I(p, \widetilde{\nu}) = I(\widetilde{\nu})[1 + cos(2\pi\widetilde{\nu}p)], \tag{1}$$

where $I(\widetilde{\nu})$ is the spectrum to be found. The total intensity at the detector is

$$I(p) = \int_0^\infty I(p, \widetilde{\nu})d\widetilde{\nu} = \int_0^\infty I(\widetilde{\nu})[1 + cos(2\pi\widetilde{\nu}p)]d\widetilde{\nu}. \tag{2}$$

This is a Fourier cosine transformation. The inverse transform can extract the desired spectrum ($I(\widetilde{\nu})$) using Equation (3). The fast Fourier transform (FFT) is used in the implementation stage. Figure 3c shows the results of spectrum extraction by applying the FFT to the interferogram (Figure 3b). The unit of the y-axis in Figure 3c is just spectral intensity in arbitrary units.

$$I(\widetilde{\nu}) = 4 \int_0^\infty [I(p) - 0.5I(p = 0)]cos(2\pi\widetilde{\nu}p)dp. \tag{3}$$

2.3. Spectral Wavenumber Calibration

A spectral or wavenumber calibration of the FTIR can be performed using the sampling theorem in digital signal processing [14]. The HYPER-CAM MWE uses a HeNe laser with wavelength $\lambda = 632.8$ nm, whose light also travels through the interferometer. The peaks of the laser signal are used to sample the received infrared (IR) source. The Nyquist sampling rate ($\tilde{\nu}_{Nyquist}$) was $1/\lambda \cdot 0.01$ (cm^{-1}) and the wavenumber spacing ($\Delta\tilde{\nu}$) was calculated by $\tilde{\nu}_{Nyquist}/N$, where the number of samples is denoted as N. In the present system, N was 1186 and $\Delta\tilde{\nu}$ was 13.3244 cycles (cm^{-1}). Therefore, the ideal wavenumber range was 0 to $13.3244 \times 1186/2 = 7901.4$ (cm^{-1}). On the other hand, the response bands of the detector (InSb) were 1807.9–6659.3 (cm^{-1}). Only 374 wavenumbers were used in the present FTIR system. Figure 3d shows the wavenumber calibrated spectrum.

2.4. Radiometric Calibration

A radiometric calibration is needed to acquire calibrated spectra in units of radiance [15]. In a space application, a blackbody (BB) and cold space can be used [16]. The HYPER-CAM MWE can provide the spectral radiance data using two BBs (hot, cold) [16,17].

The radiometric calibration involves characterizing the FTIR response by a linear equation, as expressed in Equation (4):

$$M(\tilde{\nu}) = G(\tilde{\nu}) \cdot (L(\tilde{\nu}) + O(\tilde{\nu})), \tag{4}$$

where $M(\tilde{\nu})$ is the complex spectrum from the instrument measurement, and $G(\tilde{\nu})$ and $O(\tilde{\nu})$ are the gain and offset of the instrument, respectively. $L(\tilde{\nu})$ means the true spectral radiance (W/(m$^2 \cdot$ sr $\cdot$ cm^{-1})). The gain and offset can be estimated by measuring the radiance of two known BBs. The theoretical radiance follows Planck's law, defined in Equation (5):

$$L_{BB}(\tilde{\nu}, T) = \frac{2hc^2\tilde{\nu}^3}{e^{hc\tilde{\nu}/kT} - 1}, \tag{5}$$

where $L_{BB}(\tilde{\nu}, T)$ denotes the spectral radiance (W/(m$^2 \cdot$ sr $\cdot$ cm^{-1})) of a blackbody, h is Planck's constant, c is the speed of light, k is Boltzmann's constant, and T is the blackbody temperature (K). If two BBs with known temperatures (T_H, T_C) are given, two radiances ($L(\tilde{\nu}, T_H), L(\tilde{\nu}, T_C)$) and two corresponding spectral measurements ($M_H(\tilde{\nu}), M_L(\tilde{\nu})$) are prepared. The unknown gain and offset can be obtained by solving the two equations: Equations (6) and (7). Figure 4 shows the acquired interferograms, spectra (arbitrary units), and calculated spectral radiances for the hot (95 °C) and cold (25 °C) blackbodies (first row and second row, respectively). Figure 5a,b show the estimated gain magnitude and offset magnitude at pixel (190, 10), respectively. Figure 5c presents the final estimated spectral radiance at the same pixel by comparing the data obtained by the built-in (TELOPS) calibration method. The built-in method means the spectral and radiometric calibration provided by the TELOPS instrument (Quebec, QC, Canada), which is regarded as truth. Note that similar spectral radiance can be obtained in the spectral range of 1807.9–3355.7 (cm^{-1}) using Equation (8).

$$G(\tilde{\nu}) = \frac{M_H(\tilde{\nu}) - M_C(\tilde{\nu})}{L(\tilde{\nu}, T_H) - L(\tilde{\nu}, T_C)} \tag{6}$$

$$O(\tilde{\nu}) = \frac{M_C(\tilde{\nu})L(\tilde{\nu}, T_H) - M_H(\tilde{\nu})L(\tilde{\nu}, T_C)}{M_H(\tilde{\nu}) - M_C(\tilde{\nu})} \tag{7}$$

If an unknown scene spectrum is measured ($M_S(\widetilde{v})$), the radiometrically calibrated radiance ($L_S(\widetilde{v})$) can be obtained using Equation (8):

$$L_S(\widetilde{v}) = \frac{L(\widetilde{v}, T_H) - L(\widetilde{v}, T_C)}{M_H(\widetilde{v}) - M_C(\widetilde{v})} \cdot M_S(\widetilde{v}) - \frac{M_C(\widetilde{v})L(\widetilde{v}, T_H) - M_H(\widetilde{v})L(\widetilde{v}, T_C)}{M_H(\widetilde{v}) - M_C(\widetilde{v})}. \tag{8}$$

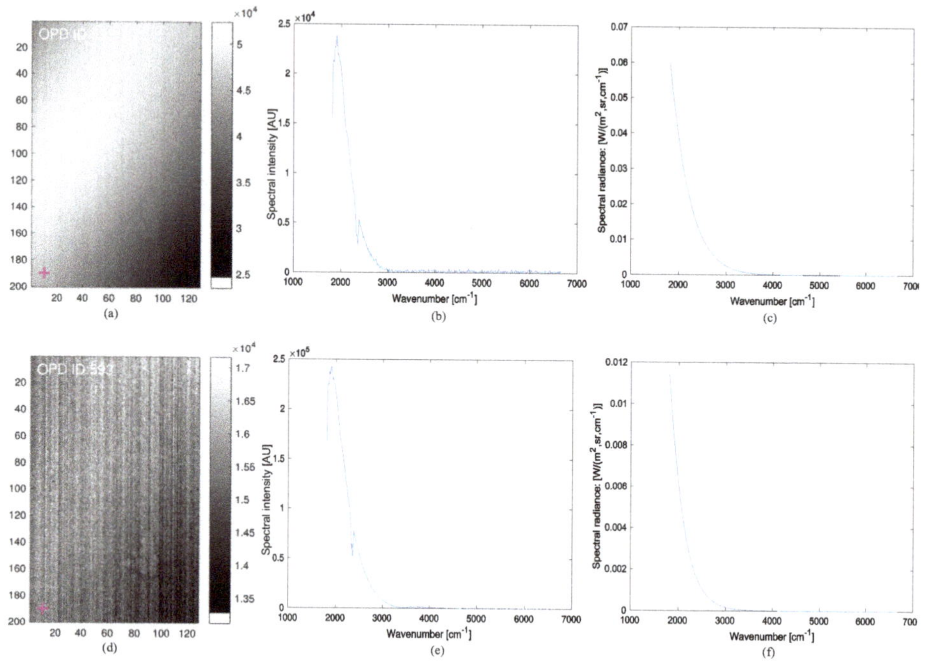

Figure 4. Blackbody spectrum and spectral radiance extraction for a radiometric calibration: (**a**) interferogram image of a hot blackbody (95 °C); (**b**) spectrum of a hot blackbody; (**c**) calculated spectral radiance at hot temperature; (**d**) interferogram image of a cold blackbody (25 °C); (**e**) spectrum of a cold blackbody; (**f**) calculated spectral radiance at cold temperature.

2.5. Brightness Temperature

The amount of spectral radiance energy can be converted into equivalent brightness temperatures [1]. By inverting Equation (5), the temperature T (K) can be obtained as

$$T = \frac{(hc/k)\widetilde{v}}{ln[2hc^2\widetilde{v}^3/L_S(\widetilde{v}) + 1]}. \tag{9}$$

Figure 5d shows the spectral brightness temperature by applying Equation (9) to the calibrated spectral radiance at pixel (190, 10). Note that there were inaccurate values due to the spectral radiance noise. The gaps in Figure 5d were generated by negative spectral radiances that were obtained during the radiometric calibration for noisy spectral radiance values. Negative spectral radiances cannot provide physical temperatures in Equation (9).

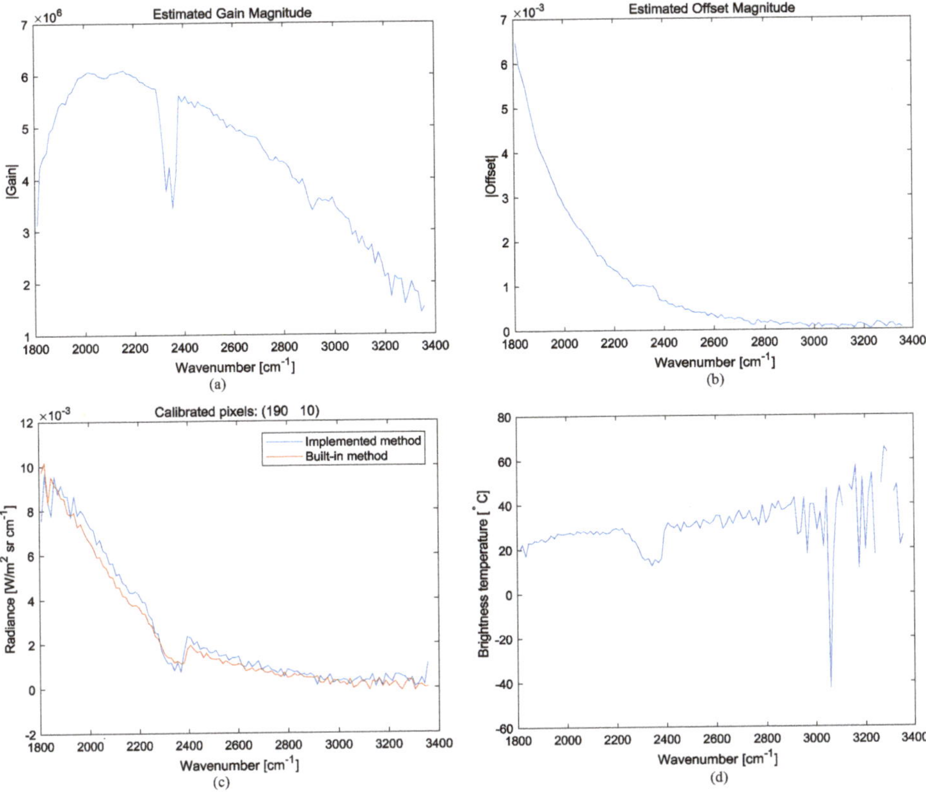

Figure 5. Radiometric calibration and brightness temperature extraction: (**a**) estimated gain magnitude; (**b**) estimated offset magnitude; (**c**) comparison of the radiometric calibration between the implemented method and built-in method; (**d**) calculated brightness temperature.

3. Proposed Temperature Estimation: ST-DCNN

An evaluation of thermal infrared stealth property is important for various applications, such as ships, cars, and houses. This paper focuses on the effects of solar radiation on objects. As shown in Figure 6 (top row), the sun heats up the surface of an object painted with specially developed materials. After 20 min (thermal equilibrium), the heated surface is rotated by 180° and radiates thermal energy (Figure 6, bottom row), which is measured by an FTIR detector. Note that the solar radiation (6000 K) is dominant in the higher wavenumber band (short wavelength region: 1.5–3.0 μm, or 3333–6667 cm^{-1}) and the energy is used to heat the surface (approximately 0–40°) of an object as indicated by the cross point. Therefore, the radiation by the heated surface is dominant in the lower wavenumber band (mid-wavelength region: 3.6–5.5 μm, or 1818–2778 cm^{-1}). The strength of the solar radiation in the shadowed region (opposite side of the direct sun) decreased to 10%–20%.

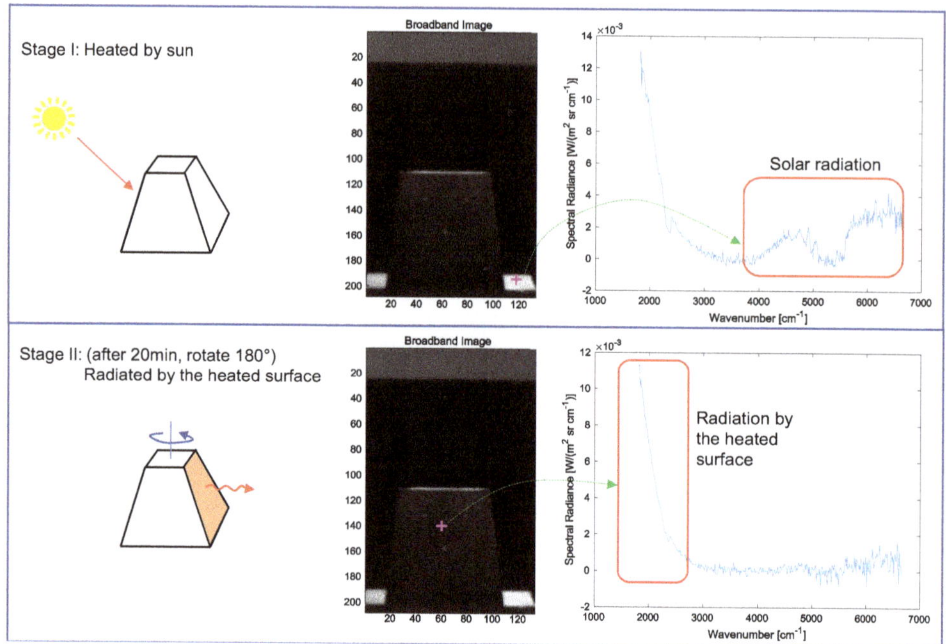

Figure 6. Thermal stealth evaluation flow caused by solar radiation at day time.

In remote surface temperature sensing, the infrared spectral radiance at the FTIR detector can be represented by radiative transfer, as per Equation (10):

$$L(\widetilde{v}) = \varepsilon_0(\widetilde{v}) L_{BB}(\widetilde{v}, T) \tau(\widetilde{v}) + L_a(\widetilde{v}), \tag{10}$$

where ε_0 is the surface spectral emissivity, $\tau(\widetilde{v})$ is the transmissivity at the FTIR detector, and $L_a(\tau(\widetilde{v}))$ represents the thermal path radiance [18]. The surface temperature (T) can be estimated from the spectral radiance ($L(\widetilde{v})$) if the object emissivity, atmospheric transmissivity, and path radiance are available. On the other hand, the environmental parameters are difficult to estimate in real-time due to the wide variations of weather conditions. Figure 7 gives an example of atmospheric transmissivity according to the object distance and weather conditions. Therefore, the one-channel method (atmospheric transmissivity is required), two-channel method (water vapor content is required), and multi-channel method (surface emissivity is required) are not applicable.

The key ideas are based on three aspects. First, a midwave thermal hyperspectral imager (HYPER-CAM MWE) is adopted to extract the spectral information. The midwave thermal hyperspectral images usually show lower sensitivity than longwave thermal hyperspectral images on the surface emissivity in a remote temperature estimation [19]. The surface emissivity can affect the temperature estimation. The low sensitivity means that the uncertainty of emissivity produces low uncertainty in temperature estimation. On a normal surface, the typical values of emissivity for MWIR are approximately 0.90–0.95. Second, the imager can acquire huge hyperspectral radiance and a temperature database at various times (10:00, 13:00, 15:00), seasons (winter, spring, summer), and weather conditions (clear, cloudy, foggy). Third, a novel surface temperature-deep convolutional neural network (ST-DCNN) was adopted to estimate the remote surface temperatures by learning the network with huge spectral radiance DB (equivalently, brightness temperature DB).

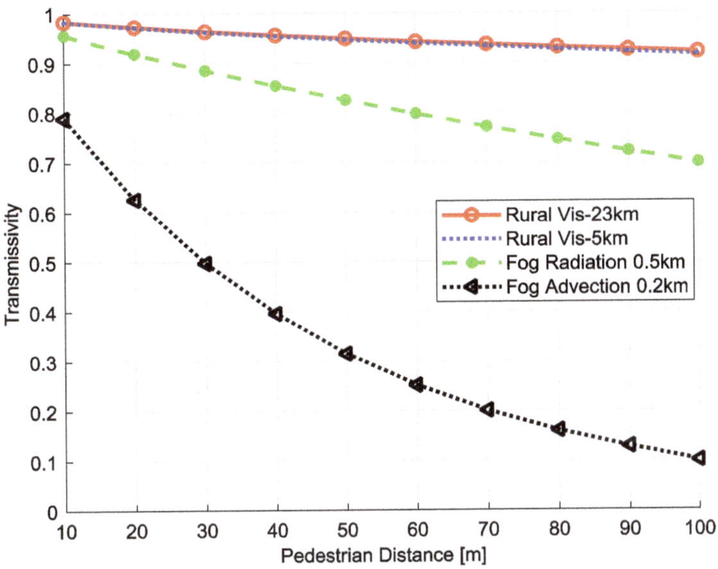

Figure 7. Atmospheric transmissivity variations according to the distance and weather conditions.

The proposed ST-DCNN consisted of 27 layers, as shown in Figure 8. The input data size was 75×1, corresponding to the midwave band (1807–2270 cm^{-1} or 3.6–5.5 µm). Two 9×1 convolutions with 64 filters, batch normalization (BN), and rectified linear unit (ReLU) were conducted to extract the temperature features in the spectral domain. The 2×1 max pooling can reduce the dimensions by removing the redundant features. Such processes (except for the last layer-1×1 Conv + BN + ReLU) were repeated twice to extract the higher spectral temperature. A 1×1 convolution was used to reduce the channel size with the same spectral feature size. The last two fully connected layers were used to regress the physical surface temperature. L2 norm was used to calculate the loss.

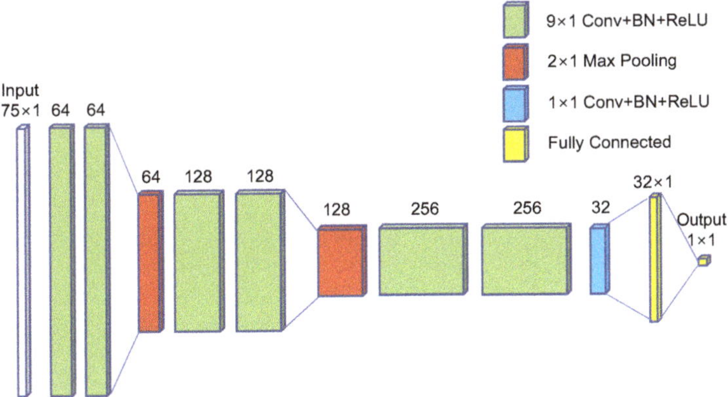

Figure 8. Proposed structure of the surface temperature-deep convolutional neural network (ST-DCNN).

4. Experimental Results

The TELOPS HYPER-CAM MWE model was used in this study, as shown in Figure 9. This model can provide a high-resolution spectrum by a Michelson interferometer (TELOPS, Quebec, QC, Canada) from the midwave to the shortwave band. The noise equivalent spectral radiance (NESR) is $7\,(\mathrm{nW}/(\mathrm{cm}^2 \cdot \mathrm{sr} \cdot \mathrm{cm}^{-1}))$ and the radiometric accuracy is approximately 2 K.

Model	Shape	Spectral Range (μm)	Spectral Resolution (cm⁻¹)	Spatial Resolution (pixel)	Field of View (deg)	NESR (nW/cm² sr cm⁻¹)	Radiometric Accuracy (K)
TELOPS HYPER-CAM MWE		1.5-5	Up to 0.25	320x256	6.5x5.1	7	2

Figure 9. Specifications of the TELOPS Hyper-Cam MWE sensor.

The object surface was painted with a gray color and was located in a coastal area with a 78 m distance to the FTIR sensor system. The object surface data were acquired from 1 December–30 June, three times a day (10:00, 13:00, 15:00) with 75 spectral bands (3.6–5.5 μm), as shown in Table 1.

Figure 10 presents an example of data preparation for deep learning. The brightness temperature data were extracted at the center region (20 × 20) and the built-in temperature sensor on the object surface provided the corresponding ground truth temperature information. Therefore, 400 brightness temperature profiles had the same physical temperature.

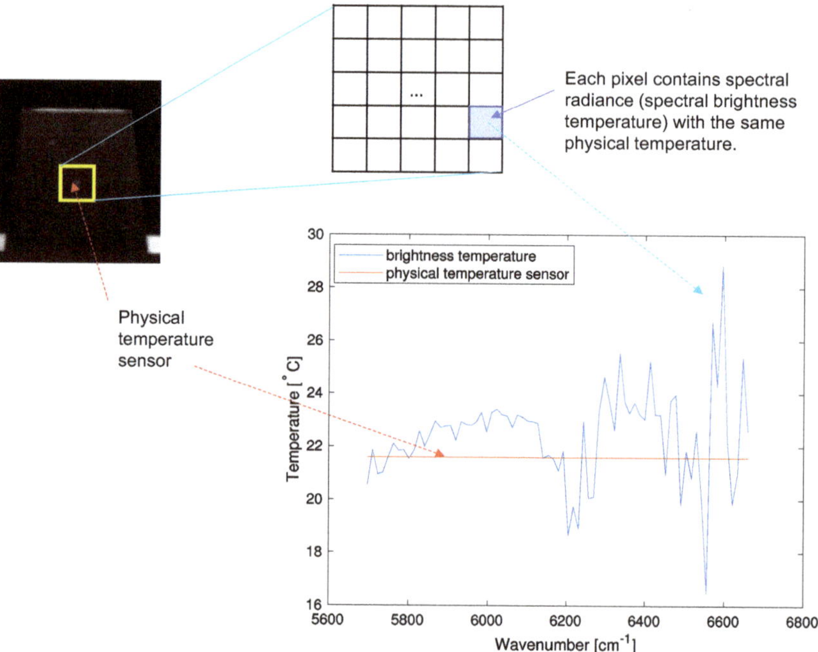

Figure 10. Preparing the spectral brightness temperature and physical temperature pair for deep learning.

Table 1. Database acquisition environment of the midwave hyperspectral spectrum.

Object	Location	Duration	Time	Spectral Band
gray painted metal	coast (78 m to sensor)	1/1–6/30	10:00, 13:00, 15:00	3.6–5.5 µm

Table 2 presents details of the spectral brightness temperature for deep learning. The number of valid hyperspectral images was 208, where each image provides 400 spectral brightness temperature profiles. The total number of spectrum–temperature pairs was 82,400, and the DB was divided into three groups: training (80%), validation (10%), and testing (10%).

Figure 11 shows the weather variations for the 208 hyperspectral images acquired from 1 December 2017–30 June 2018 in terms of the air temperature, relative humidity, and solar radiation. The air temperature ranged from -10 to 28 °C, the relative humidity varied from 20% to 100%, and the solar radiation varied from 0 to 1000 W/m^2 .

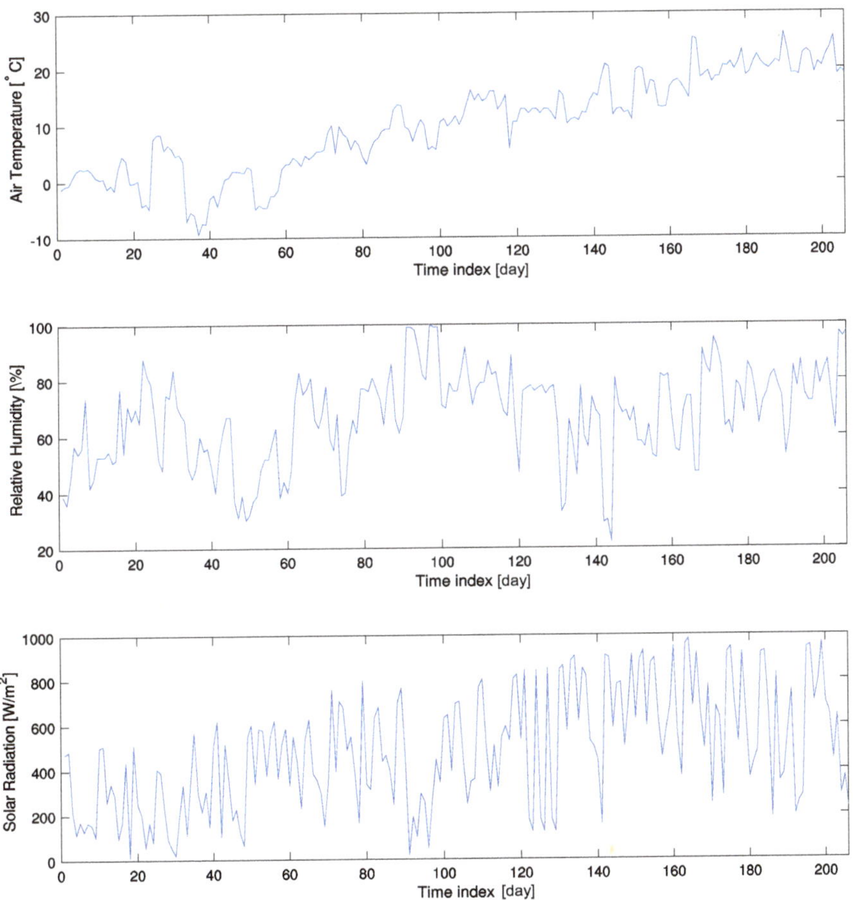

Figure 11. Preparing the spectral brightness temperature and physical temperature pairs for deep learning.

Table 2. Composition of the spectral brightness temperature data for deep learning.

No. of Valid Images	Crop Size	No. of Training (80%)	No. of Validation (10%)	No. of Testing (10%)
208	20×20	65,920	8240	8240

The proposed ST-DCNN consisted of 27 layers that were optimized to produce the best regression performance, as shown in Table 3. The number of pooling layers was two, where the max pooling showed better performance than the average pooling. The VGG-style convolutions with the kernel size 9 showed the best root mean square error (RMSE) among the kernel sizes (3, 5, 7, 9, 11, and 13). Additional batch normalization was helpful, and drop-out was useless. A 1×1 convolution was used to reduce the feature dimension, and the regression performance was upgraded if the 1×1 convolution was inserted at the end of the last Conv–BN–ReLU stage. The best performance was seen with 32 fully connected layers, among 8, 16, 32, 64, and 128.

Figure 12 shows the training process with the optimized network parameters. Approximately seven minutes were needed to train the ST-DCNN with a mini-batch size of 128, max epochs of 30, initial learning rate of 1.0×10^{-4}, learning rate drop factor of 0.2, learning rate drop period of 20, and shuffle in every epoch.

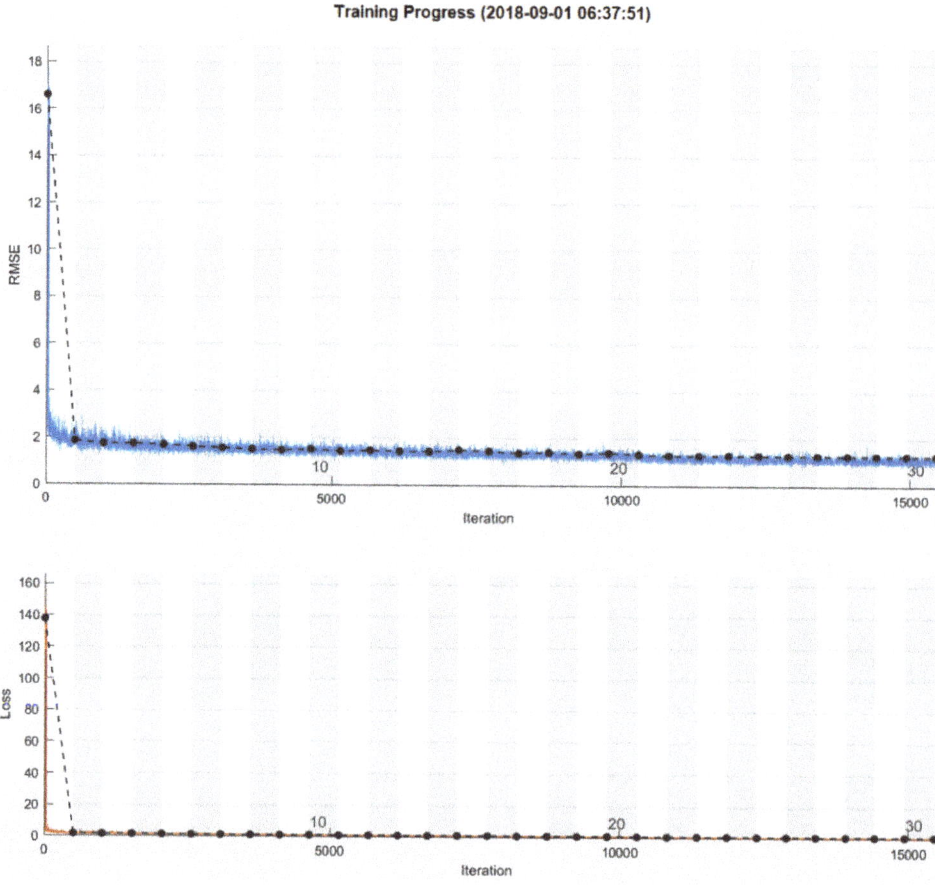

Figure 12. Training process: (**top**) root mean square error curve; (**bottom**) loss curve.

Table 3. Optimization results of deep learning parameters. RMSE: root mean square error.

No. of pooling layers	1	**2**	3	4	-
RMSE	1.33	**1.14**	1.26	1.28	-
No. of filters	16–32–64	32–64–128	**64–128–256**	128–256–512	256–512–1024
RMSE	1.29	1.23	**1.14**	1.29	1.37
Filter size	5	7	**9**	11	13
RMSE	1.27	1.22	**1.14**	1.19	1.23
No. of nodes in FC	8	16	**32**	64	128
RMSE	1.19	1.18	**1.14**	1.21	1.23
Dropout rate	0.2	0.3	0.4	0.5	**W/O**
RMSE	1.21	1.21	1.21	1.29	**1.14**
1×1 conv. size	16	**32**	64	128	W/O
RMSE	1.23	**1.14**	1.22	1.22	1.25
Batch norm. (BN)	**W/**			W/O	
RMSE	**1.14**			1.77	

The proposed ST-DCNN-based temperature estimation method was compared with the multi-layer perceptron (MLP) [20] and mean brightness temperature [19]. The MLP consisted of two hidden layers with 128 nodes. Figure 13 shows the partial results tested on the unlearned 8240 samples. Figure 13b is an enlarged graph of Figure 13a. Note that the proposed ST-DCNN could predict the true temperatures better than the other methods. Table 4 lists the quantitative comparison results in terms of the RMSE measures. The RMSE value of the direct temperature estimation using the mean brightness temperature was 2.0934 °C and that of the MLP was 1.7863 °C. The proposed ST-DCNN showed an RMSE of 1.1446 °C. This was improved by 45.32% compared to the base method (mean brightness temperature (BT)). Note that the RMSE of the brightness temperature method was similar to the manufacturer's radiometric accuracy (2 K). Although the RMSE of the proposed ST-DCNN was approximately 1.14 °C, it was reasonably accurate considering the wide weather variations along the three seasons, as shown in Figure 11. Figure 14 presents the remote temperature estimation results for March 29, 15:00 DB. The ground truth temperature of the center region was 19.8 °C, and the proposed ST-DCNN predicted correctly. On the other hand, the MLP and mean BT methods estimated incorrectly by 1 °C.

Table 4. Comparison of the temperature estimation in terms of the RMSE.

Method	RMSE (°C)	Improvement (%)
MLP [20]	1.7863	14.67
Proposed (ST-DCNN)	**1.1446**	**45.32**
Brightness temperature [19]	2.0934	0
TELOPS MWE: radiometric accuracy	2	-

One of many applications of remote temperature estimation is to check the thermal stealth effect of different types of paint on ships, buildings, etc. by solar radiation. Figure 15 shows a quantitative visualization of the physical surface temperature on an object. The surface temperature was estimated using the proposed ST-DCNN with the best network parameters. Figure 15a represents a broadband image of the experimental environment imaged on 5 March 2018. Figure 15b,c show the temperature distributions of Plate-C and Plate-D, respectively. Each plate was painted with different types of surface material. The weather conditions at 15:40 were as follows: atmospheric temperature 10.2 °C, humidity 55%, and solar radiation 550 W/m^2. The estimated average temperature of Plate-C was 14.7 °C, whereas that of Plate-D was 12.7 °C. Therefore, the paint on Plate-D had 2.0 °C better thermal stealth capability. In addition, the physical temperature distribution of Plate-C and Plate-D could be compared.

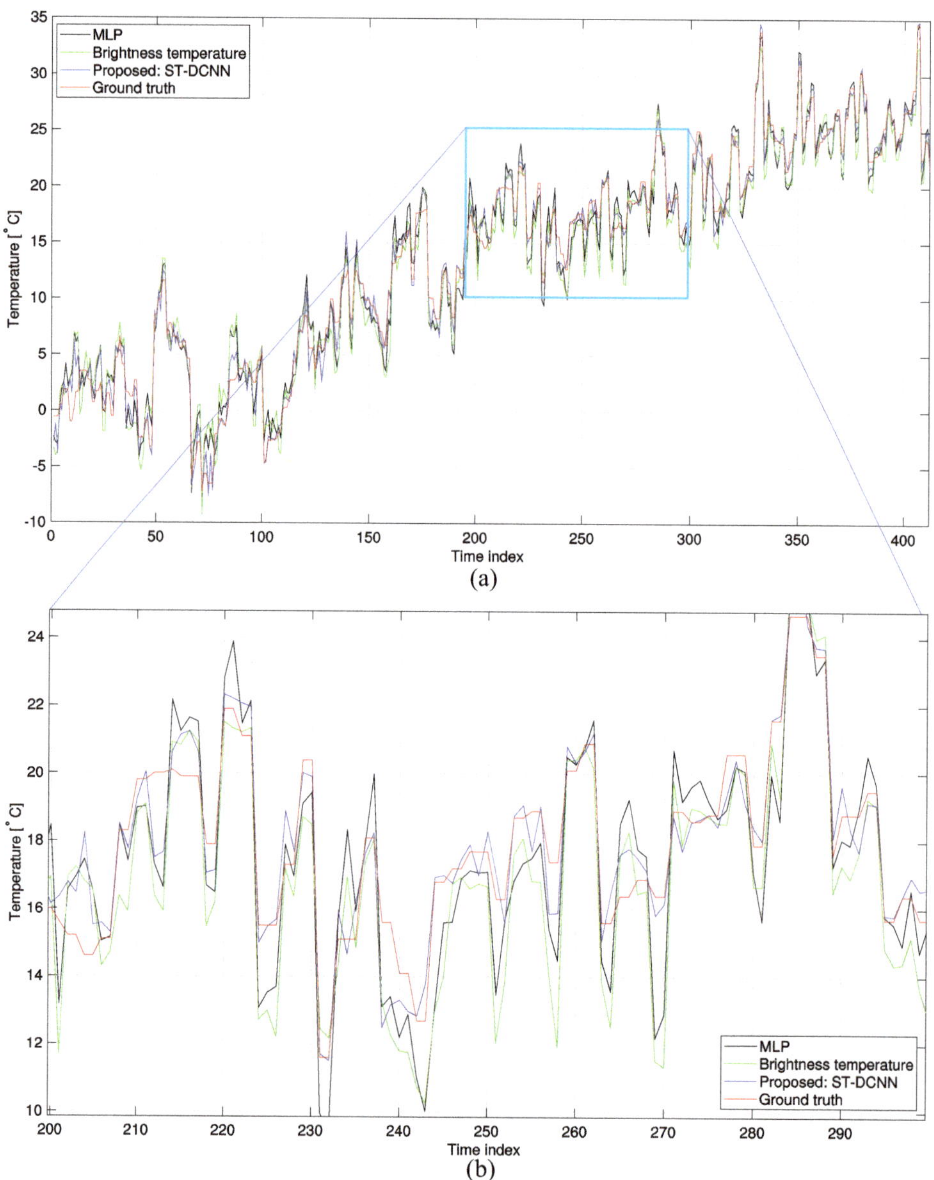

Figure 13. Comparison of the temperature estimation: (**a**) overall temperature; (**b**) enlarged view of the probe region. MLP: multi-layer perceptron.

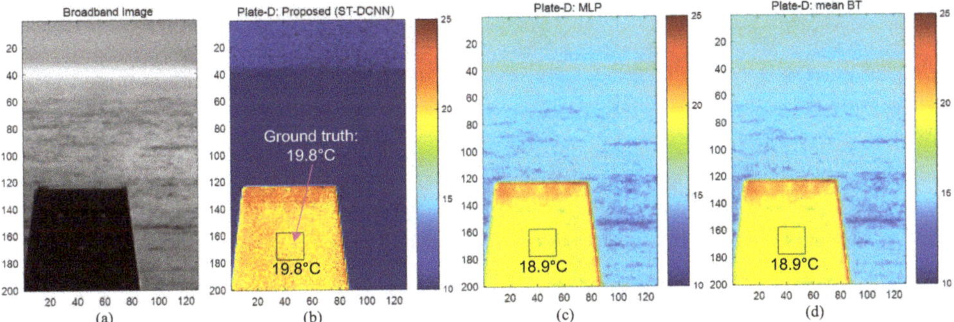

Figure 14. Visualized comparison of temperature estimation: (**a**) broadband image; (**b**) proposed method (ST-DCNN); (**c**) MLP; and (**d**) mean brightness temperature (BT).

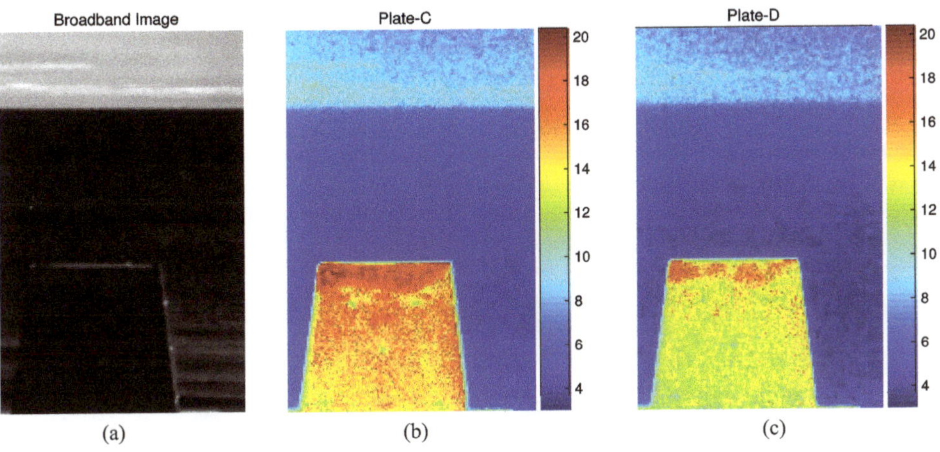

Figure 15. Thermal stealth application of a remote temperature estimation: (**a**) broadband image; (**b**) temperature distribution of Plate-C; (**c**) temperature distribution of Plate-D.

5. Conclusions and Further Works

This paper proposed a novel remote surface temperature estimation method using a surface temperature-deep convolutional neural network (ST-DCNN) from midwave hyperspectral thermal images. Estimating the surface temperature remotely in a dynamically varying weather environment is a very challenging problem. The key idea was to apply the specially designed 27-layer deep convolutional neural network to a huge database consisting of spectral radiance–surface temperature sensor pairs. The spectral radiance data were extracted successfully by applying the FFT to the interferogram followed by wavenumber calibration and two blackbody-based radiometric calibrations. The inverse of Planck's law to the spectral radiance produced the spectral brightness temperature containing both the surface emissivity and atmospheric transmissivity implicitly. The proposed ST-DCNN learned the weight parameters successfully using the 65,920 training samples. According to the experimental results, the proposed ST-DCNN showed an RMSE of 1.1446 °C, which is 45% better than the average brightness temperature. The remote temperature estimation scheme was applied to evaluate the thermal stealth effects by the solar radiance on different types of paint. In the future, a multi-sensor fusion-based deep learning structure will be developed to improve the accuracy of temperature estimation.

Author Contributions: The contributions were distributed between authors as follows: S.K. wrote the text of the manuscript and programmed the hyperspectral temperature estimation method using ST-DCNN. J.K., J.L. and J.A. provided the midwave infrared hyperspectral database, operational scenario, performed the in-depth discussion of the related literature, and confirmed the accuracy experiments that are exclusive to this paper.

Funding: Basic research programs funded by ADD, Yeugnam university, and NRF.

Acknowledgments: The research was supported by the Agency for Defense Development (UE17104GD). This research was supported by the 2018 Yeungnam University Research Grants. This research was also supported by Basic Science Research Program through the National Research Foundation of Korea(NRF) funded by the Ministry of Education (NRF-2018R1D1A3B07049069).

Conflicts of Interest: The authors declare no conflict of interest.

References

1. Trishchenko, A.P. Solar Irradiance and Effective Brightness Temperature for SWIR Channels of AVHRR/NOAA and GOES Imagers. *J. Atmos. Ocean. Technol.* **2006**, *23*, 198–210. [CrossRef]

2. Wubet, M.T. Estimation of Absolute Surface Temperature by Satellite Remote Sensing. Ph.D. Thesis, International Institute for Geoinformation and Earth Observation, Enschede, The Netherlands, 2003.

3. Tonooka, H. Accurate atmospheric correction of ASTER thermal infrared imagery using the WVS method. *IEEE Trans. Geosci. Remote Sens.* **2005**, *43*, 2778–2792. [CrossRef]

4. Hook, S.; Gabell, A.; Green, A.; Kealy, P. A comparison of techniques for extracting emissivity information from thermal infrared data for geologic studies. *Remote Sens. Environ.* **1992**, *42*, 123–135. [CrossRef]

5. Li, H.; Liu, Q.H.; Du, Y.M.; Jiang, J.X.; Wang, H.S. Evaluation of the NCEP and MODIS atmospheric products for single channel land surface temperature retrieval with ground measurements: A case study of HJ-1B IRS data. *IEEE J. Sel. Top. Appl. Earth Observ. Remote Sens.* **2013**, *6*, 1399–1408. [CrossRef]

6. McMillin, L.M. Estimation of sea surface temperatures from two infrared window measurements with different absorption. *J. Geophys. Res.* **1975**, *80*, 5113–5117. [CrossRef]

7. Sun, D.; Pinker, R.T. Retrieval of surface temperature from the msg-seviri observations: Part I. Methodology. *Int. J. Remote Sens.* **2007**, *28*, 5255–5272. [CrossRef]

8. Zhong, X.; Labed, J.; Zhou, G.; Shao, K.; Li, Z.L. A Multi-Channel Method for Retrieving Surface Temperature for High-Emissivity Surfaces from Hyperspectral Thermal Infrared Images. *Sensors* **2015**, *15*, 13406–13423. [CrossRef] [PubMed]

9. Zhong, X.; Huo, X.; Ren, C.; Labed, J.; Li, Z.L. Retrieving Land Surface Temperature from Hyperspectral Thermal Infrared Data Using a Multi-Channel Method. *Sensors* **2016**, *16*, 687. [CrossRef] [PubMed]

10. Beil, A.; Daum, R.; Matz, G.; Harig, R. Remote sensing of atmospheric pollution by passive FTIR spectrometry. *Proc. SPIE* **1998**, *3493*, 32–43.

11. Roy, S.A. Data Processing Pipelines Tailored for Imaging Fourier-Transform Spectrometers. Ph.D. Thesis, Université Laval QuéBec, Québec, QC, Canada, 2008.

12. Griffiths, P.R.; de Haseth, J.A. *Fourier Transform Infrared Spectroscopy*; John Wiley and Sons: Hoboken, NJ, USA, 1986.

13. Atkins, P.; Paulah, J.D. *Physical Chemistry*, 8th ed.; Oxford University Press: Oxford, UK, 2006.

14. Skoog, D.A.; Holler, F.J.; Crouch, S.R. *Principles of Instrumental Analysis*, 6th ed.; Thomson: Belmont, CA, USA, 2007.

15. Hook, S.J.; Kahle, A.B. The micro Fourier Transform Interferometer (mu FTIR)—A new field spectrometer for acquisition of infrared data of natural surfaces. *Remote Sens. Environ.* **1996**, *56*, 172–181. [CrossRef]

16. Worden, H.; Beer, R.; Bowman, K.W.; Fisher, B.; Luo, M.; Rider, D.; Sarkissian, E.; Tremblay, D.; Zong, J. TES Level 1 Algorithms: Interferogram Processing, Geolocation, Radiometric, and Spectral Calibration. *IEEE Trans. Geosci. Remote Sens.* **2006**, *44*, 1288–1296. [CrossRef]

17. Schlerf, M.; Rock, G.; Lagueux, P.; Ronellenfitsch, F.; Gerhards, M.; Hoffmann, L.; Udelhoven, T. A Hyperspectral Thermal Infrared Imaging Instrument for Natural Resources Applications. *Remote Sens.* **2012**, *4*, 3995–4009. [CrossRef]

18. Sun, D.; Pinker, R.T. Estimation of land surface temperature from a Geostationary Operational Environmental Satellite (GOES-8). *J. Geophys. Res.* **2003**, *108*, 4326. [CrossRef]
19. Qian, Y.G.; Zhao, E.Y.; Gao, C.; Wang, N.; Ma, L. Land Surface Temperature Retrieval Using Nighttime Mid-Infrared Channels Data From Airborne Hyperspectral Scanner. *IEEE J. Sel. Top. Appl. Earth Observ. Remote Sens.* **2015**, *8*, 1208–1216. [CrossRef]
20. Wang, N.; Tang, Z.L.B.H.; Li, F.Z.C. Retrieval of atmospheric and land surface parameters from satellite-based thermal infrared hyperspectral data using a neural network technique. *Int. J. Remote Sens.* **2013**, *34*, 3485–3502. [CrossRef]

MDPI

St. Alban-Anlage 66

4052 Basel

Switzerland

Tel. +41 61 683 77 34

Fax +41 61 302 89 18

www.mdpi.com

Micromachines Editorial Office

E-mail: micromachines@mdpi.com

www.mdpi.com/journal/micromachines

9 783036 501741